Nicole Lindenmann

Photonic Wire Bonding as a Novel Technology for Photonic Chip Interfaces

Karlsruhe Series in Photonics & Communications, Vol. 21
Edited by Profs. C. Koos, W. Freude and S. Randel

Karlsruhe Institute of Technology (KIT)
Institute of Photonics and Quantum Electronics (IPQ)
Germany

Photonic Wire Bonding as a Novel Technology for Photonic Chip Interfaces

by
Nicole Lindenmann

KIT Scientific Publishing

Karlsruher Institut für Technologie
Institut für Photonik und Quantenelektronik

Photonic Wire Bonding as a Novel Technology
for Photonic Chip Interfaces

Zur Erlangung des akademischen Grades eines Doktor-Ingenieurs
von der KIT-Fakultät für Elektrotechnik und Informationstechnik des
Karlsruher Instituts für Technologie (KIT) genehmigte Dissertation

von Dipl.-Phys. Nicole Lindenmann geboren in Pforzheim

Tag der mündlichen Prüfung: 27. April 2017
Referent: Prof. Dr. Christian Koos
Korreferenten: Prof. Dr. Uli Lemmer, Prof. Dr. Dr. h. c. Wolfgang Freude

Impressum

KIT Scientific Publishing

Karlsruher Institut für Technologie (KIT)
KIT Scientific Publishing
Straße am Forum 2
D-76131 Karlsruhe

KIT Scientific Publishing is a registered trademark
of Karlsruhe Institute of Technology.
Reprint using the book cover is not allowed.

www.ksp.kit.edu

Print on Demand 2018 – Gedruckt auf FSC-zertifiziertem Papier

ISSN 1865-1100
ISBN 978-3-7315-0746-8
DOI 10.5445/KSP/1000077963

To my family

"Courage ist gut. Ausdauer ist besser."

Theodor Fontane

Table of Contents

List of Figures

Kurzfassung

Die Übertragung großer Datenmengen ist heute so alltäglich wie nie zuvor, und das weltweite Datenaufkommen wächst nach wie vor ungebremst. Der „Cisco Visual Networking Forecast 2014–2019" sagt ein Anwachsen des globalen Datenverkehrs auf über 1 Zettabyte im Jahr 2016 voraus. Milliarden von Gigabytes an Daten werden täglich über das Internet verschickt, in Datenzentren gespeichert und bearbeitet. Während die Datenübertragung dabei auf optischem Wege erfolgt, werden die Daten innerhalb der Datenzentren vorwiegend elektronisch verarbeitet und gespeichert. Daher werden leistungsfähige Schnittstellen zwischen den optischen Übertragungswegen und den elektronischen Speicher- und Verarbeitungseinheiten benötigt. Da in Großrechneranlagen mehrere Zehntausend dieser Schnittstellen erforderlich sind, ist für deren Herstellung nur eine Technologie sinnvoll, die es erlaubt, kompakte, energieeffiziente optische Transceiver in großen Stückzahlen kostengünstig herzustellen. Die integrierte Photonik, bei der optische, wellenleiterbasierte Komponenten planar auf Halbleiter-Chips realisiert werden, erfüllt diese Anforderungen an geringem Platzbedarf und skalierbaren Stückzahlen. Insbesondere die Silizium-Photonik, die auf ausgereifte CMOS-Prozesse zurückgreifen kann, wird dabei als eine sehr zukunftsfähige Technologie angesehen. Im Moment ist es insbesondere die Aufbau- und Verbindungstechnik dieser integriert-optischen Systeme, die noch immer einen Flaschenhals bei der Realisierung und Verbreitung der Technologie darstellt. Um erfolgreich industriell verwertbar zu sein, muss die integrierte Photonik durch eine Aufbau- und Verbindungstechnik ergänzt werden, mit der kompakte Schnittstellen preisgünstig und zuverlässig hergestellt werden können.

Leistungsfähige optische Einzelmodule und erste Systeme können dank enormer technologischer Fortschritte auf verschiedenen Materialplattformen hergestellt werden. Das Fehlen einer idealen Aufbau- und Verbindungstechnik hat jedoch auch zur Folge, dass viele optische Module und Systeme gar nicht realisiert werden können. Obwohl die einzelnen Bausteine wie Laser, Modulatoren und Detektoren in einer Vielzahl von Materialsystemen vorhanden sind, können die nötigen Einzelkomponenten oft nicht auf ein und derselben Materialplattform hergestellt werden. Eine skalierbare Aufbau- und Verbindungstechnik, die die verschiedenen Materialplattformen verbindet und ihre jeweiligen Vorteile in geeigneter Weise kombinieren kann, könnte es hingegen ermöglichen, optische Multi-Chip-Systeme mit bisher unerreichter Funktionalität und Leistungsfähigkeit zu realisieren.

Im Moment werden zur optischen Verbindung von Einzelkomponenten hauptsächlich Standard-Einmodenfasern verwendet. Die chip-basierten Bauteile werden direkt mit diesen Fasern verbunden. Dieses Verfahren benötigt viel Platz, da der minimale Abstand zwischen zwei optischen Ankoppelstellen auf dem Chip durch den mit 125 µm sehr großen Durchmesser der Glasfaser diktiert ist. Ein weiterer Nachteil besteht darin, dass sich die Modenfelder der stark führenden optischen Wellenleiter auf dem Chip massiv vom Modenfeld der Glasfaser unterscheiden. Zusätzliche, zum Teil sehr lange Wellenleitersegmente sind notwendig, um die Fehlanpassung der Modenfelder zu beheben. Das verbraucht weitere wertvolle Chip-Fläche. Der wohl größte Nachteil bei der direkten Verbindung zwischen Fasern und integrierten photonischen Komponenten besteht darin, dass ein serielles aktives Ausrichten (active alignment) der Einzelkomponenten mit höchster mechanischer Präzision notwendig ist. Die einzelnen Bauteile werden dabei im aktiven Betrieb mit Genauigkeiten von unter 500 nm relativ zu einander positioniert, während der Kopplungsverlust zwischen Chip und Glasfaser permanent gemessen und durch Relativbewegung der Komponenten optimiert wird. Dieses Verfahren ist technisch sehr anspruchsvoll und inhärent seriell, was den erreichbaren Durchsatz limitiert. Vor allem bei integrierten photonischen Komponenten, die keine eigene Laserquelle aufweisen, treten große technische Schwierigkeiten auf.

Die vorliegende Arbeit beschreibt ein neuartiges Verfahren zur Erzeugung optischer Verbindungswellenleiter, das sich im Gegensatz zum aktiven Ausrichten einzelner Bauteile durch ein hohes Maß an Flexibilität und Automatisierbarkeit auszeichnet. Dabei werden polymere Wellenleiter, sog. photonische Wirebonds, direkt an die Koppelstellen von grob vorpositionierten, integriert-optischen Komponenten angesetzt. Photonisches Wirebonden beruht auf der Erzeugung dieser dreidimensionalen optischen Verbindungswellenleiter mit direkt-schreibender Laserlithographie. Diese neue Technologie hat dabei einige entscheidende Vorteile gegenüber existierenden Verfahren, wie zum Beispiel das aktive Ausrichten. Die durch photonisches Wirebonden entstehenden Verbindungswellenleiter sind sehr kompakt. Der Abstand zwischen einzelnen Verbindungsstellen kann wenige Mikrometer betragen, so dass keine zusätzliche Chipfläche für ein Auffächern der Wellenleiter verbraucht wird. Eine Anpassung des Modenquerschnitts zwischen den zu verbindenden Wellenleitern kann durch die Formgebung des photonischen Wirebonds erfolgen und eine Anpassung des Wellenleiterquerschnitts auf dem Chip ist nicht mehr zwingend erforderlich. Photonisches Wirebonden ermöglicht es insbesondere, die Vorteile verschiedener Materialsysteme miteinander zu kombinieren, um optische Multi-Chip-Systeme mit bisher unerreichten Leistungsmerkmalen zu schaffen. Diese Systeme können in verschiedenen photonischen Anwendungsfeldern von der Datenkommunikation, über die Messtechnik bis hin zur Biophotonik zum Einsatz kommen. Eine hochpräzise Relativpositionierung verschiedener Komponenten ist dabei nicht nötig, da die einzelnen photonischen Wirebonds dynamisch an die realen Positionen der einzelnen Komponenten angepasst werden. Die Technologie lässt sich weitestgehend automatisieren und zu hohem Durchsatz skalieren.

Die vorliegende Arbeit beschreibt die Simulation, technische Realisierung und experimentelle Validierung von photonischen Wirebonds. Sie beinhaltet die Demonstration von Schlüsselelementen, wie die Verbindung zwischen silizium-basierten Wellenleitern, die Anbindung von aktiven optischen Komponenten, sowie die Verbindung zwischen Glasfasern und Silizium-Chips.

Summary

Cloud services, ubiquitous network access and "Internet of Things" applications all rely on powerful global communication networks. "Cisco Visual Networking Index Forecast" (2014–2019) predicts that global IP traffic will transcend the zettabyte threshold in 2016 by reaching a monthly traffic of 88.4 Exabyte. Those data packages are transmitted over the network, and are stored and processed in large data centers. While the transmission *via* global networks and inside the data centers predominantly relies on optical techniques, data storage and processing are mainly done electronically. For this reason efficient interconnects between optical transmission links and electronic storage and processing units are required in large counts. A technology that can provide compact, energy efficient optical transceivers fabricated in large volume at low cost is therefore needed. Highly integrated photonic integrated circuits (PIC) fabricated from semiconductors using scalable micro- and nanofabrication techniques can provide the required device count and the necessary miniaturization.

Driven by substantial worldwide research investments, tremendous progress has been made in the field of photonic integration over the last years. Both continuous advances of III-V material systems and the advent of silicon photonics that exploits mature CMOS technology, have enabled highly scalable on-chip integration of photonic transceiver systems. However, there is currently no technology that could cope with the associated photonic inter-chip connectivity challenges. Currently this is one of the main obstacles towards industrial adoption of large-scale photonic-electronic integration.

Moreover, the lack of a scalable interconnect technology also hinders the advance of novel system architectures that combine the strength of different integration platforms in a heterogeneous approach. Photonics is currently characterized by the co-existence of several material platforms, all having specific advantages and weaknesses. Importantly, there is no platform that can provide all functionalities simultaneously. To overcome this drawback, it is of essential importance to create a scalable optical interconnect technology that can combine the strength of different optical integration platforms enabling multi-chip photonic systems with highest efficiency and functionality.

In conventional photonic systems, devices are connected by standard single mode fibers. As a consequence, the minimum achievable interconnect density is dictated by the cross-section of the fiber, which is often much larger than the geometrical footprint of the on-chip component. A further drawback of direct fiber-to-chip coupling is the severe mismatch between the optical mode of the strongly-guiding on-chip waveguides and of the optical fiber. To adapt the mode fields, long spot-size converters need to be added to the on-chip waveguides which consumes precious chip space. The most severe disadvantage of direct fiber-to-chip coupling, however, is the need for active alignment. In the case of active alignment optical components are adjusted with respect to each other, while the optical transmission through the assembly is measured and optimized. The mechanical tolerances need to be smaller than 500 nm, which makes expensive machinery indispensable. The process is serial and time-consuming and thus inherently limited in throughput. In addition to that, this expensive and complex technology cannot be easily applied to photonic components that do not possess an own light source, as it is the case in silicon photonic devices.

These deficiencies can be overcome by photonic wire bonding. Here three-dimensional (3D) polymer waveguides are fabricated directly between the interfaces of waveguides. Photonic wire bonds (PWB) are very compact and thus give an advantage over conventional coupling approaches. The accessible pitch can be as small as several micrometers and on-chip waveguide fan-outs are therefore not necessary. In addition, mode field adaption can be performed inside the PWB, thus rendering long dedicated on-chip tapers unnecessary and further reducing the footprint of the interface. The three-dimensional shape of the photonic wire bond can be adapted to the exact positions of the feed waveguides. Tolerances in mechanical chip placement can be compensated for by the shape of the PWB, this makes the use of high-precision mechanical alignment in chip positioning obsolete. Photonic wire bonding can be used for both chip-to-chip and fiber-to-chip connections and enables low-loss broadband transmission. The technique lends itself to an automated assembly of photonic systems. It offers the potential to connect known-good dies of different materials to form high-performance multi-chip systems, thereby exploiting the complementary strengths of the various integration platforms.

In the framework of this thesis, we developed the concept of photonic wire bonding, we established the processes and created the necessary tools. We, finally, demonstrated all key elements, required for the fabrication of photonic multi-chip systems. Those are photonic wire bonds between passive optical components, between passive and active devices, as well as between on-chip components and optical fibers.

This thesis provides the summary of the fundamental concept. It gives the design considerations, and reports on the experimental proof-of-principle demonstrations. The individual chapters of this work cover the following subjects:

Chapter 1 gives an introduction to the field of photonic integration and multi-chip systems. It further provides an overview of the photonic wire bonding concept including the principles of multi-photon polymerization. The associated lithography setup and the different experimental setups used in this work are discussed.

Chapter 2 describes the photonic wire bonding workflow in detail. It explains how photonic wire bonds can be adapted to the feed waveguides. The detection of the interface positions and the trajectory planning of the PWB are explained. The generation of the 3D PWB trajectory and shape is introduced and different writing strategies together with their strengths and weaknesses are discussed. The work presented in this chapter is covered by two international patents [P2], [P3].

Chapter 3 deals with the experimental realization of photonic wire bonds connecting silicon-on-insulator (SOI) waveguides across chip boundaries. Simulation results and proof-of-principle demonstrations are given for chip-to-chip connections, their losses are evaluated, and data-transmission capabilities are analyzed. The results presented in this chapter led to various publications including an article in "Optics Express" [J2] and multiple conference contributions [C6], [C8], [C15], [C16], [C20], [C21].

Chapter 4 addresses the coupling of PWB to active optical components *via* vertical device facets. Mode field adaption is discussed and simulations of the interface are provided. In this configuration, the lithography beam is partially disturbed by the presence of the laser's vertical device facet. Based on a theoretical model, a simulation environment is set up that allows analyzing the influence of shadowing edges on the lithography laser beam. From these simulations, compensation curves are derived and experimentally verified. Finally, a proof-of-principle device fabrication and characterization is demonstrated.

Chapter 5 reports on the first demonstration of fiber-to-chip connections by photonic wire bonds. The chapter discusses the simulation of the coupling interfaces, as well as the simulation of the influence of the fiber facet to the lithography laser beam during fabrication. An experimental verification with test structures proves the simulation results. Fiber-to-fiber as well as fiber-to-chip PWB were successfully realized and are presented in this chapter. The chapter further discusses the coupling between photonic wire bonds and multicore fibers. The experimental results presented in this chapter led to an article in the "IEEE Journal of Lightwave Technology" [J1] as well as to a conference contribution [C6].

Chapter 7 summarizes the presented work and gives suggestions for future directions.

Achievements of the Present Work

In the framework of this thesis, we developed photonic wire bonding as an interconnection technology for multi-chip photonic systems, including all necessary tools, algorithms, and processes. The viability of the concept has been examined and proven by numerous experiments.

Process Development for Photonic Wire Bonding:

In this work we developed a complete fabrication procedure resulting in smooth, low-loss photonic wire bonds (PWB). The process exploits a self-developed on-chip marker recognition that can be used to precisely detect the waveguide facets in the coordinates of the lithography system. We have implemented a three-dimensional (3D) routing algorithm to define a suitable waveguide trajectory between two desired waveguides ports. This algorithm avoids sharp bends and unnecessary long paths for low transmission loss. Around the PWB trajectory, we create a 3D bond structure that incorporates dedicated taper structures for optimized coupling to and from different waveguides. Whereas Chapter 1 provides the fundamentals of photonic wire bonding, Chapter 2 focuses in detail on the developed process steps and algorithms that are used to create PWB adapted to specific sample geometries. The technology development on photonic wire bonding forms the basis of two international patents [P2], [P3].

Fabrication of PWB Connections between Silicon-on-Insulator Chips

Photonic wire bonds enable low-loss broadband connections to silicon photonic circuits. In proof-of-concept experiments, losses down to 1.1 dB were achieved. Error-free data transmission of up to 5.25 Tbit/s has been demonstrated. The results have in part been published in [J2], [C21], [C15], [C8] and are discussed in depth in Chapter 3.

Fabrication of PWB Connections between Lasers and Silicon-on-Insulator Chips

Overcoming one of the major drawbacks in silicon photonics, which is the lack of a laser source, we demonstrated the connection of a III-V-laser to a silicon-on-insulator waveguide *via* a PWB. Coupling to edge emitting semiconductor lasers by photonic wire bonding is inherently challenging since the vertical device facet blocks part of the lithography beam. In the framework of this thesis, we derived a theoretical model and set up a simulation environment to analyze and thus compensate the perturbations introduced by a vertical device facet. Using this compensation technique, a successful proof-of-principle devices fabrication for a laser-to-chip PWB is presented in Chapter 4.

Fabrication of PWB Connections between Fibers and Silicon-on-Insulator Chips

Whereas interconnections between single chips with the help of standard single mode fiber (SMF) can become obsolete using PWB, coupling photonic multi-chip systems to the outside world still requires a reliable connection to the standard single mode fiber. Therefore a main achievement of the presented work is the interconnection between SMF and SOI. In addition to that, the fabrication of a single-mode photonic wire bond interconnection between SOI waveguides and the individual cores of a four-core fiber with excess losses as low as 1.7 dB is demonstrated in Chapter 5. Parts of this chapter have been published in [J1].

International Recognition

The work on photonic wire bonding presented in this thesis has in parts been disseminated in numerous publications and has been highly received by the international research community: The first achievements have been accepted for presentation as a post-deadline paper at the "Optical Fiber Communication Conference" (OFC) in 2011 [C21], followed by a publication in "Optics Express" [J2] and a series of international conference presentations [C8], [C20], [C16]. Moreover, the topic led to a series of invited talks at international conferences [C15], [C11]-[C14], [C9], [C7], [C5], [C3]. The research on PWB for fiber-to-chip coupling was first presented at the IEEE Optical Interconnects (OI) Conference in May 2014 [C6] and was extended into an article in IEEE Journal of Lightwave Technology [J1] in a special issue on Optical Interconnects. Regarding industrial exploitation, the work covered in this thesis has led to two international patents [P2], [P3] and one national patent application [P1] and has formed the basis for a national research project [1]. Moreover, the concept is used in different European research projects, aiming at energy-efficient terabit/s transceivers. The feedback of the involved industry partners confirms the notion that the concept of photonic wire bonding has the potential to revolutionize integrated photonics by enabling multi-chip systems that combine the specific advantages of different integration platforms.

1 Introduction: Photonic Multi-Chip Integration and Photonic Wire Bonding

1.1 State of the Art in Optical Input/Output-Technology

Integrated optics and in particular silicon photonics have been the subject of intense research and development efforts in recent years. Silicon photonics is based on the integration of photonic components on silicon-on-insulator (SOI) substrates. Silicon photonic technology permits transferring mature complementary metal oxide semiconductor (CMOS) processes developed for mass production of integrated electronic circuits to photonic integration. SOI devices based on transverse single-mode, planar integrated waveguides[1] provide strong optical confinement, enable light transport around narrow bends and hence lead to a high integration density. Nano-photonic systems with high complexity and a plurality of functionalities can thus be compactly and efficiently produced on an industrial scale. The areas of application for such systems are primarily data transmission, but also the fields of optical metrology and sensing. The economic potential of silicon photonics is huge, and first products are already commercially available [2]. The high density of on-chip integration in silicon photonics, however, must be complemented by an interconnect technology with comparable scalability and versatility for achieving large-scale industrial adoption.

Previous research in the field of optical input/output technologies has primarily focused on refining fiber-to-chip coupling. Out-of-plane coupling has been demonstrated using diffraction gratings etched into the top surface of SOI

[1] Waveguide structures are called "transverse single mode" if for each polarization only the basic mode can propagate. Transverse single-mode waveguides thus generally have two guided modes with different polarizations, for example called "quasi-TE" and "quasi-TM". A low-loss optical connection of these systems is therefore only possible with waveguide structures which permit an efficient connection to the basic modes of the integrated waveguide.

waveguides [3]–[11]. When properly designed, such structures can simultaneously act as couplers and integrated polarization splitters [7]. The coupling efficiency can be improved by using bottom mirrors [10], by silicon overlays [5], or by using numerically optimized gratings [6], leading to insertion losses between 0.62 dB and 2 dB for advanced devices [12]. Grating couplers enable optical access anywhere on the chip surface, but 1 dB-bandwidths are limited to typical values between 40 nm and 50 nm. To achieve broadband transmission, it is necessary to use in-plane coupling schemes that combine inverse tapers [13] with polymer [14], silicon dioxide (SiO_2) or silicon oxynitride (SiO_xN_y) waveguides [15]–[17]. With such structures, coupling losses of less than 1 dB for a single fiber-to-chip interface can be maintained over wavelength ranges of more than 100 nm [16]. However, in this approach the spatial interconnect density is still limited by the pitch of the connecting fibers. Advanced multi-channel spot-size converters allow to interface an array of single-mode fibers with a pitch of 250 µm to an array of SOI waveguides with a 20 µm pitch, thereby enabling spatial data transmission density of 8 Tbit/s per millimeter chip edge [18]. Scaling to high numbers of input/output waveguides, however, results in increasingly complex fabrication processes for the spot-size converters. Moreover, all state-of-the-art off-chip interconnect techniques rely on the highly precise positioning of external fibers with respect to on-chip coupling structures. Fabrication is mostly based on intricate active alignment techniques, where the coupling efficiency is dynamically monitored and optimized, and often involves manual fabrication steps. Active alignment is especially difficult to implement in the case of passive photonic components that do not feature an integrated laser source, such as silicon-on-insulator waveguides. As on-chip integration density increases further, the restrictions of costly fabrication and limited interconnect density will become even more pronounced.

Monolithic integration is often presented as a method to avoid costly chip-to-chip interfaces in photonic systems, but the processing complexity and the performance limitations are still an obstacle for many applications. Photonics is still characterized by a co-existence of several material platforms with complementary strengths: Silicon photonics uses foundry-based high-yield CMOS processes for high-density integration of passive devices along with electro-optic modulators and germanium-based photo detectors [19], [20]. III-V compound semiconductors such as Indium Phosphide (InP) are the mainstay for highly efficient integrated optical sources, amplifiers, as well as for high-speed electro-optic modulators and photo detectors [21], [22]. Waveguides made of silicon oxide, silicon nitride or from polymers feature low and medium index contrast, and are commonly used to realize optical filters and other high-performance passive components [23], [24]. An industrially viable technology for chip-to-chip interconnects could enable novel system concepts that efficiently combine the strengths of different integration platforms, without giving up the many advantages of sophisticated on-chip integration.

1.2 The Photonic Wire Bonding Concept

In the field of microelectronics, the development of a reliable metal wire bonding technique was a basic requirement for economic success. Methods with comparable throughput and degree of automation have so far not been available for the field of photonics.

In direct analogy to electronics, where metal wire bonds connect integrated circuits, transparent, free-standing polymer waveguides are used in this work to bridge the gap between nano-photonic circuits located on different chips or different photonic components. The 3D freeform polymer waveguides are *in situ* generated between the coupling locations of feed waveguides. These polymer waveguides are referred to as photonic wire bonds (PWB) in the following. The underlying fabrication process exploits three-dimensional direct laser writing (3D DLW) based on multi-photon induced polymerization. In the approach followed in this work, the shapes of the PWB are adapted to the actual positions of the integrated waveguide facets. For this reason, high-precision alignment of optical devices becomes obsolete. The photonic wire bonding concept allows for very compact interconnects between different components. Interconnect pitches down to several micrometers are easily achievable. Since photonic wire bonding is based on 3D laser lithography, the shape as well as the cross-section of the PWB can be adapted to the feed waveguides. Spot-size converters can be integrated into the PWB, thus, ultra-long on-chip tapers optimized for fiber-coupling and expensive consumption of chip-space are not required any more. Photonic wire bonding enables co-integration of different photonic device technologies in flexible multi-chip systems. A schematic illustration of a multi-chip system can be seen in Fig. 1.1. The technique allows for automated assembly and mass production of photonic systems and is therefore expected to be of great importance for the industrial deployment of photonic integrated circuits.

The fabrication of PWB comprises several fabrication steps. An illustration of the photonic wire bonding concept for the fabrication of photonic multi-chip systems is depicted in Fig. 1.2.

In a first step the photonic components are mounted on a common sub-mount. This sub-mount can be a planar microscope cover glass in the simplest case. In the case of more advanced modules that comprise active components, metallic sub-mounts can be chosen. The submount then acts simultaneously as a mechanical ground plate, a heat sink, an electrical contact as well as a device for rough height adjustment. Bare dies are mounted on the sub-mount using polymer adhesives, conductive adhesives, or soldering technologies. In the mounting procedure, only rough alignment of the devices is required with an accuracy easily accessible by standard pick-and-place machinery.

An upper limit for the distance between the outlets of the two feed waveguide is given by the maximum coverage area of the lithography system. This range amounts to 300 μm×300 μm×300 μm for the first machine generation used in this thesis. The more recent machine generation provides a larger writing field of 500 μm×500 μm×300 μm.

In a second process step, the module including sub-mount and photonic components is cleaned by successive rinsing with ultra-clean acetone and isopropyl alcohol to remove organic remains as well as loose inorganic particles.

After sample preparation, the optical module together with the resist is inserted into the 3D direct laser writing lithography system.

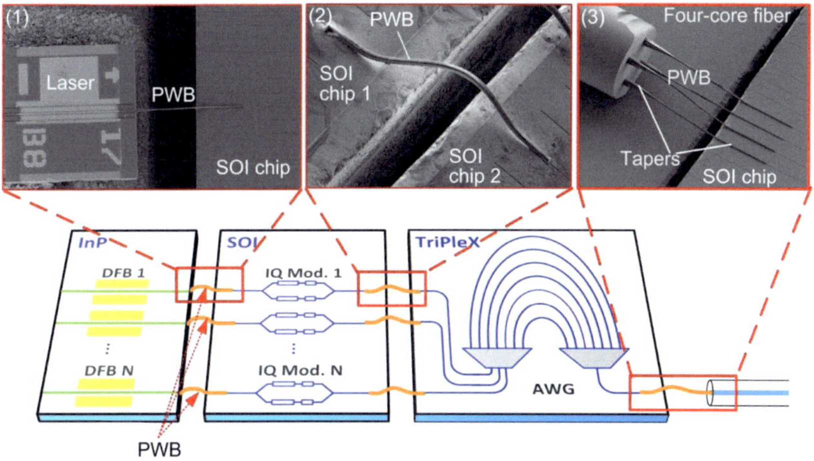

Fig. 1.1: Illustration of photonic wire bonding as enabling technology for multi-chip photonic systems. Photonic wire bonds are used as interconnect waveguides between different photonic material platforms. For illustration, the example of a transmitter for wavelength-division multiplexing (WDM) communications is depicted. The system combines the specific strengths of different material systems: Distributed-feedback lasers (DFB 1 … N) are realized on direct-bandgap InP-substrates and optically coupled to electro-optic modulators (Mod 1 … N) on a silicon-on-insulator (SOI) chip. High-performance arrayed waveguide gratings (AWG) are needed for dense packing of optical channels. These devices can be realized in medium index-contrast material systems such as TriPleX, a commercially available platform that enables low-loss waveguides based on a combination of silicon nitride (Si_3N_4) and silicon dioxide (SiO_2) [23]. The feasibility of high-performance multi-chip systems is vitally dependent on low-loss chip-to-chip and chip-to-fiber connections, which can be efficiently realized by photonic wire bonds (depicted in orange). **Inset (1):** Laser-to-chip connection: a photonic wire bond is used to interconnect a DFB laser to an SOI waveguide. **Inset (2):** Chip-to-chip connections: Photonic wire bond connecting two SOI waveguides on adjacent chips. **Inset (3):** Fiber-to-chip connections: Photonic wire bonds connecting the individual cores of a multi-core fiber (MCF) to an array of planar SOI waveguides.

In a third step, photo resist material is deposited between the interconnect regions of the feed waveguides. Depending on the resist material used, this third step differs for cationic and radical resists. The different experimental setups for both types of resists are discussed in Section 1.3.2.

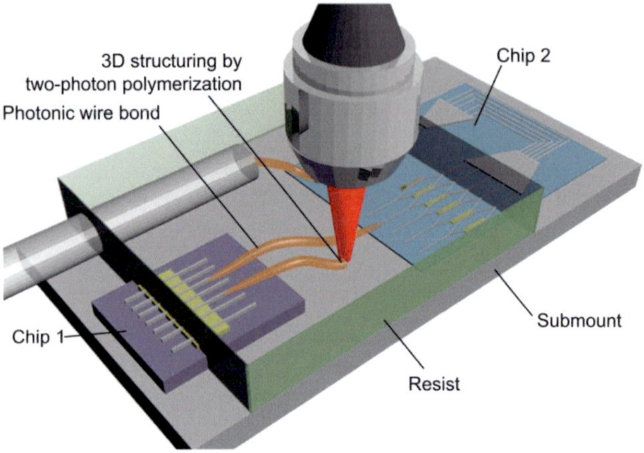

Fig. 1.2: Artist's impression of the photonic wire bonding concept, modified from [J2]: Photonic chips (Chip 1, Chip 2) and optical fiber ends are mounted on a carrier and embedded into a photosensitive negative-tone photo resist. After measuring the relative positions of the integrated waveguide facets, matched PWB geometries are computed. The waveguide structures are then defined by a 3D direct laser writing technique which is based on two-photon polymerization in the focus of a pulsed laser beam. The unexposed regions are removed in a development step and can subsequently be clad with a low-index cover material.

A camera is used to obtain a visual impression of the sample. Using a pattern recognition tool that we developed in the framework of this thesis, optical images obtained by the camera are used to measure the positions of the optical feed-waveguide. This measurement has to be very accurate (tolerances below 100 nm) to ensure correct positioning of the photonic wire bonds with respect to the waveguide outlets. More details about the image-based pattern recognition will be given in Section 2.1.

In a next step, the shapes of the photonic wire bonds are calculated based on the measured coordinates of the waveguide outlets. Section 2.2 gives a detailed description of the trajectory planning, whereas Sections 3.2.3, 4.1, and 5.1 discuss the mode field adaption used to generate low loss photonic wire bonds connecting several different devices. Since the waveguide geometry is adapted to the relative positions of the photonic subsystems, no high-precision active alignment is required during fabrication, which makes the technique a good candidate for the fully automated industrial production of optical modules.

In order to generate a machine readable dataset, the calculated structure of the PWB has to be transferred into single writing lines consisting of single points that are consecutively addressed by the lithography system. Depending on the resist material and the desired shape of the photonic wire bond, different methods can be used to generate this dataset. These methods will be referred to as "writing strategies" in the following and are discussed in detail in Section 2.4.

In the lithography step, the pre-calculated structure of the PWB is generated using the aforementioned writing strategy. Afterward, the non-polymerized resist material is removed in a developer bath.

In an optional last processing step, the developed PWB can be clad with a low-index cover material that acts as protection as well as a mechanical support.

The following sections give an overview of the fundamentals of multi-photon induced polymerization, the used lithography system, as well as the sample preparation for photonic wire bonding.

1.3 Base Technology of Photonic Wire Bonding

The underlying fabrication technique exploited in photonic wire bonding is 3D direct laser writing based on multi-photon induced polymerization [25]–[27]. The following section discusses the fundamentals of multi-photon polymerization, the different types of sample setups required for different resist materials, and the lithography systems used in this thesis.

1.3.1 Multi-Photon-Induced Polymerization

"Polymers are macromolecules built up by the linking together of large numbers of much smaller molecules. The small molecules that combine with each other to form polymer molecules are termed *monomers*, and the reactions by which they combine are termed *polymerizations*." [28]

Multi-photon induced polymerization is a light-induced polymerization process. It is triggered by simultaneous absorption of multiple photons. In the case of two-photon induced polymerization, the cross-linking process of the resist requires two photons to be absorbed simultaneously. A mathematical expression for the two-photon absorption rate p in dependence of the electrical field strength $\vec{E}$ is given by [29]. It depends on the factor $\mathrm{Im}\left\{\chi^{(3)}\right\}$, the imaginary part of the nonlinear optical susceptibility $\chi^{(3)}$ of third order in the form

$$p \propto \mathrm{Im}\left\{\chi^{(3)}\right\} |\vec{E}|^2 \cdot |\vec{E}|^2 . \tag{1.1}$$

Far away from resonances in the material, a simplified case can be considered where the complex third-order nonlinear susceptibility $\chi^{(3)}$ tensor can be approximated by a complex scalar. The absorption rate p is high, i. e., the two-photon induced polymerization is effective, if the photon density is very high, i. e., if the squared optical field $|\vec{E}|^2$ is very strong. To guarantee these high field strengths, setups for 3D DLW usually employ strong, pumped lasers with fs pulse-width and repetition rates in the MHz range. The described nonlinearity of the absorption rate can be used to spatially confine the polymerization to a small area around the geometrical focus of the lithography system. This is most efficiently reached by the use of high numerical aperture objectives to strongly focus the lithography beam. The small 3D region around the focus of the lithography system that undergoes polymerization in the two-photon polymerization will be referred to as "voxel" in the following. By moving the focal point of the lithography system through the volume of a negative-tone photo resist material[2], along a trajectory in three dimensions, almost any arbitrary 3D structure can be created.

While this mechanism is used here for the generation of optical waveguides, the applications of multi-photon induced polymerization range from photonics where the technique is used for the fabrication of photonic crystals [27], [30], to surface chemistry using micro-scaffolds [31] as well as biology and tissue engineering [32].

3D DLW based on multi-photon polymerization has also been used for the fabrication of optical waveguides in numerous publications [33]–[36]. Whereas the work presented here relies on free-standing three-dimensional freeform waveguides, the structures fabricated in previous studies rely on a local increase of the refractive index that is induced by multi-photon polymerization in special resist material such as Ormocer® [37]. In those studies, areas with increased refractive index are used as a waveguide core. The resulting structures remain embedded in the resist material that is hardened in a post-processing step and that acts as a lower refractive optical cladding material. The achievable index contrast in this method, however, is low (typically $\Delta = \left(n_1^2 - n_2^2\right) / 2n_1^2 = 0.005$). This results in rather large waveguides. The realization of compact chip-to-chip connections, and especially the adaption to strongly guiding photonic integrated circuits, as it is studied in this thesis, is not possible with those techniques.

The work in this thesis predominantly uses commercially available negative-tone resist material. These resists contain at least a photo-initiator, and a pre-polymer component such as a monomer or oligomer that forms larger chains or clusters during the polymerization process or during the cross-linking reaction. Among the commercially available photo resists there are two major

[2] A negative-tone photo resist is a resist material in which unexposed regions can be removed in a solvent bath. The exposed regions undergo polymerization, they are solidified, and withstand the solvent. A positive-tone photo resist is a resist material in which the unexposed regions withstand the solvent, whereas the exposed regions can be dissolved by the solvent.

groups that differ in the polymerization or cross-linking reaction: Cationic and radical photo resists [26].

In a **cationic** photo resist, the photo-initiator releases an ion upon illumination. That ion triggers the polymerization or cross-linking chain reaction. Usually, this reaction only takes place in a so-called 'post exposure bake' step at increased temperatures that enhance the diffusion of ions into the photo resist. A typical representative of this group of photo resists is Epon SU-8 [38] that has also been used in the framework of this thesis.

In a **radical** photo resist, the photo-initiator releases a free radical instead of an ion. This radical then starts the polymerization process. An additional 'post exposure bake' is not required. In the framework of this thesis several radical photo resists have been tested and used, amongst these *IP-Dip*™ [39] has proven to be a good candidate for the fabrication of photonic wire bonds. It provides fast fabrication due to easy sample preparation and post-treatment.

1.3.2 The Photonic Wire Bonding Setup

Depending on the resist type used for photonic wire bonding, the sample preparation and sample setups differ. In the framework of this thesis different resist materials have been evaluated. In this section the two sample setups for cationic as well as for radical resist materials will be discussed

Cationic photo resists require an elaborate sample preparation and post treatment. As an example the sample preparation for SU-8 [38] is given:

After mounting the single photonic components onto a common carrier and a cleaning step of the module, the cationic resist material is spin coated or drop cast onto the sample. The module together with the resist is then covered by a microscope cover glass that serves as an optical interface towards the lithography system. The resist material is thus sandwiched between the optical module and the microscope cover glass, as it is displayed in Fig. 1.3. For SU-8 a pre-exposure bake is required that transforms the liquid resist into a gel-like or even solid layer, since solvents in the resist are evaporated in this step. This pre-bake is performed at typical temperatures of 65°C and is applied for 5 minutes, followed by a 45-minute bake at 95°C. The duration of the second baking step, however, depends on the layer thickness of SU-8. The formation of a solid or gel-like layer provides some advantages for the following lithography step. Since SU-8 [38] provides a solid resist layer, isolated writing lines do not move during the lithography process, as it is the case in a liquid resist. This offers a greatly simplified writing strategy since no special precautions have to be taken to "anchor" every writing line to an existing or pre-fabricated solid structure. In addition, the refractive index change of the material upon illumination is negligible, thus writing through structures that have been already exposed does not result in any voxel degradation [40]. The sandwiched sample preparation required for cationic photo resists, however, can influence the lithography parameters in an undesired way.

In the case of a cationic resist material, sample developing after the lithography step also includes a post-exposure bake. In this step the actual polymerization takes place. The post-exposure bake is followed by a developer bath.

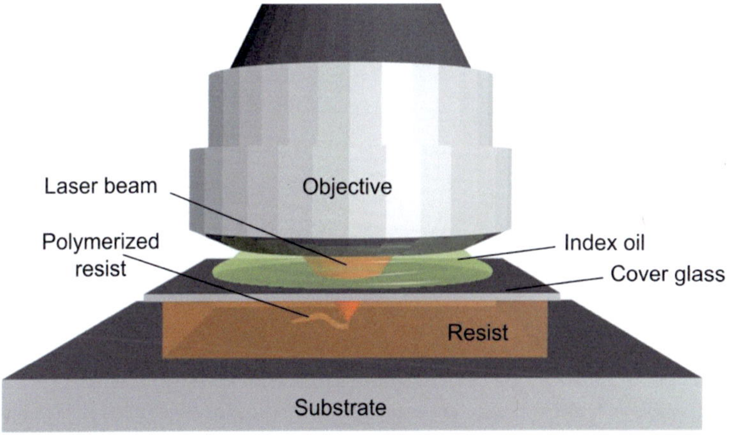

Fig. 1.3: Schematic illustration of the sample setup for cationic resists. The substrate is coated with the resist. The module is then covered by a microscope cover glass that serves as an optical interface towards the lithography system. The resist material is thus sandwiched between the optical module and the microscope cover glass. Index oil is used as an immersion liquid between the lithography objective and the cover glass.

Compared to the cationic resist, sample preparation in the case of a **radical** resist is rather easy, as it does not require a pre-bake. In the case of *IP-Dip*™[39] the resist is drop-cast on the module. A microscope cover is not necessary in this case since the lithography objective can be brought into direct contact with the resist, as can be seen in Fig. 1.4. This resist, however, does not form a solid or gel-like matrix which requires additional attention in the selection of an adequate writing strategy. A key advantage of this resist is that the optical path length inside the resist is kept constant at all times during lithography since the objective is directly immersed in the used resist. An additional interface in form of a cover glass is not required, and therefore additional optical aberration effects can be avoided.

In the case of a radical resist like *IP-Dip*™ [39] the polymerization already takes place in the lithography step, and a post-exposure bake after the lithography step is not required. Like in the cationic case, the non-polymerized resist is removed in a developer bath.

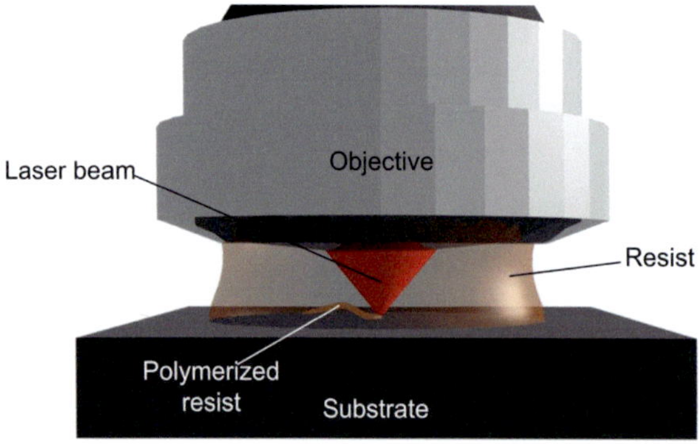

Fig. 1.4: Schematic illustration of the sample setup for "Dip-In" laser lithography [41]: A microscope cover is not necessary in this case since the lithography objective can be brought into direct contact with the resist without the need for an additional optical interface. The non-polymerized resist material stays liquid during the lithography process.

1.3.3 The Two-Photon-Polymerization Lithography Setup

In the framework of this thesis several versions of the commercially available system Photonic Professional® by Nanoscribe [39] have been used. The basic system comprises a femtosecond laser[3] system at a central wavelength of $\lambda = 780\,\mathrm{nm}$. Depending on the setup chosen the repetition rates vary slightly around typical values of $f_{\mathrm{rep}} = 80\,\mathrm{MHz}$. The laser system emits femtosecond pulses at typical pulse durations of $\tau < 100\,\mathrm{fs}$. The optical power that enters the lithography objective is set by a computer controlled acousto-optic modulator (AOM). A microscope serves as the opto-mechanical platform of the system that carries a high numerical aperture (NA) objective used to tightly focus the lithography laser beam. Typical numerical apertures amount to $\mathrm{NA} = n \sin\alpha = 1.3$. Here the refractive index n is dictated by the immersion liquid, and the angle α describes the half opening angle under which the laser beam is focused. Movement of the sample with respect to the fixed focus position of the system is accomplished *via* a three-axis stage with piezo transducers. This piezo stage is addressed by the process control software *via* a computer interface and moves the sample with respect to the focus of the lithography system. Since in this version of the setup the sample is moved with respect to a fixed focus position the inertia of the mechanically moving parts is rather high and lead to a limited accessible writing velocity v. The system comprises further a less precise sample stage for rough positioning. A

[3] Direct laser writing in general exploits a nonlinear process that causes the focal volume to polymerize while the off-focal volume remains solvable. It has to be mentioned, however, that in the more general case of direct laser writing, the nonlinearity does not have to be caused by the optical absorption process. The nonlinearity can also arise from other effects, such as oxygen-scavenging free radicals [101], [102]. In these cases the nonlinearity is caused by the chemistry of the involved resist itself. Exploiting this direct laser writing has been shown using inexpensive cw-laser sources [103], [104]. The vast majority of direct laser writing systems, however, exploit multi-photon induced polymerization and use femtosecond laser pulses at wavelength around 800 nm. Femto-second lasers provide sufficient peak intensity to drive an efficient multi-photon polymerization at low average laser power [105].

high-resolution camera is used to monitor the process. This camera system is used for image-based sample recognition and position measurement.

In a new version of the system (Photonic Professional GT®, [39]) the focus of the lithography beam is moved with respect to a fixed sample. The lateral movement of the laser beam is accomplished by galvanometer-driven scanning mirrors in a setup similar to that depicted in Fig. 1.5. The galvanometer-driven mirrors change the incident angle into the lithography objective. This change in the angle is transferred into a lateral movement of the focus in the sample plane. The axial direction is still covered by movement of the sample closer to and further away from the objective. This novel technique allows much higher writing velocities v perpendicularly to the optical axis and thus reduced fabrication times for structures that can easily be divided into slices.

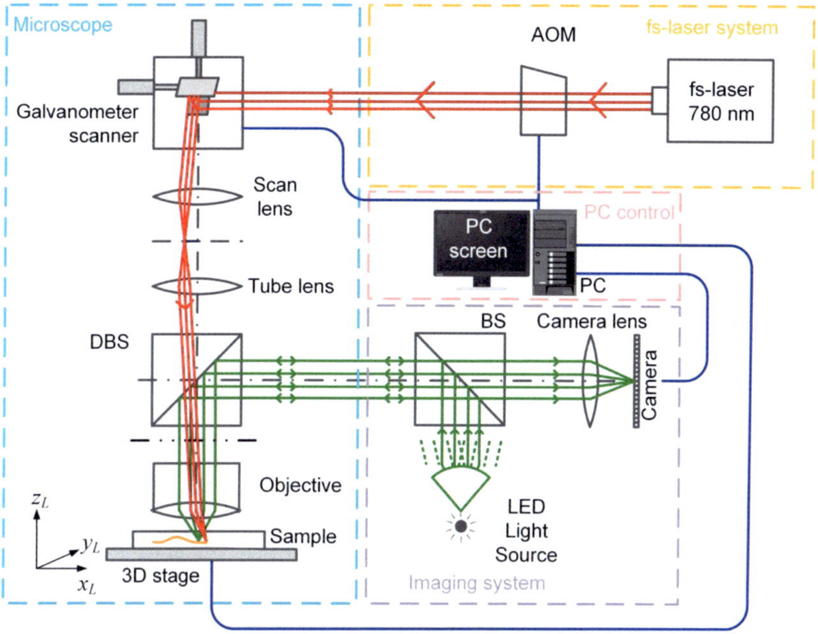

Fig. 1.5: Schematic illustration of a galvanometer mirror based lithography system. Light from a fs-laser system is sent through an acousto-optical modulator (AOM). Depending on the applied voltage, the modulator diffracts a portion of the incident light to higher diffraction orders. Only light from one selected diffraction order can pass through the system, the laser power that reaches the sample space can thus be controlled by the AOM. Inside the microscope, galvanometer mirrors are used to change the incident angle of the laser beam into the lithography objective. A scan and tube lens serve to translate the angle-offsets introduced by the galvanometer mirrors into a lateral offset in the sample space. The sample is fixed to a three-axis piezo stage. The stage is used for coarse positioning of the sample as well as for the fine axial movement of the sample with respect to the focal point during lithography. An imaging system is used to provide a live overview of the sample as well as for marker detection and recognition, here a light-emmiting diode (LED) provides the illumination of the system. The camera and the LED light path are combined by a beam splitter (BS). The illumination light and the lithography laser beam are combined in a dichroic beam splitter (DBS). The whole set-up is controlled by a PC that runs the machine control software. The path of the lithography light is depicted in red, the illumination light path is shown in green, and blue lines indicate electrical interconnections between devices and the PC control system.

2 Photonic Wire Bonding Workflow

In order to obtain low interconnect loss photonic wire bonds have to be adapted to the connecting photonic components. First of all, photonic wire bonds have to be aligned to the desired coupling position with highest accuracy (better than 100 nm). Second, the photonic wire bond trajectory has to be chosen, such that losses due to sharp bends are avoided. The cross-section of the photonic wire bond along the trajectory has to be chosen to optimally match the mode fields of the interconnected components. Finally, the surface of the photonic wire bond has to be smooth to minimize scattering loss. For high surface quality, the resist material, as well as the writing strategy have to be carefully chosen.

The following section describes the individual steps of the photonic wire bonding process in more detail.

2.1 Port Detection Algorithm

To ensure optimum coupling between the photonic wire bonds and the photonic components to be interconnected, a precise knowledge of the coupling positions as well as a reliable alignment of the photonic wire bonds with respect to these positions is of crucial importance. In order to reliably detect the positions and directions of on-chip waveguides, a marker-based port detection technique has been developed in the framework of this thesis. This became necessary since the Photonic Professional® does, at the current stage, not provide an inbuilt calibration of camera coordinate system with respect to the lithography coordinate system, nor an inbuilt marker detection scheme. In the following we refer to the optical axis of the lithography system as the z_L-axis, the x_L- and y_L-axis of the lithography system are given by the axis of the piezo transducers. These three axes form a right-handed orthonormal system, as can be seen in Fig. 2.1.a.

Our algorithm makes use of the camera built into the Photonic Professional® system and works in two stages:

1. Tilt correction:

Even under optimized experimental conditions any sample will be tilted with respect to the z_L-axis of the lithography system to a certain extent. In a first stage of the port detection algorithm the tilt of the sample with respect to the optical axis of the lithography tool is measured[1]. The measurement of this tilt relies on the fact that the camera and the lithography beam of the Photonic Professional® are aligned to share either a common focal plane at height z_L or two focal planes with a constant and stable offset z_{LC}. Let us assume a well-aligned system in the following, which means that the plane in which two-photon polymerization takes place is also the plane that results in the best focused camera pictures. It is further assumed that the sample surface is planar and the z-axis of the piezo system is perpendicular to the x-y-plane of the camera.

For the tilt measurement and correction, stacks of images corresponding to different z-heights are taken. In order to obtain a sufficient amount of data, the z-stacks are collected at least at four predefined positions $P_1...P_4 = (x_1, y_1)...(x_4, y_4)$ in the lithography or piezo x-y-plane, as displayed in Fig. 2.1.a. This is done by moving the sample to the predefined positions and collecting images at different z-positions. Out of these (minimum four) image stacks, the z-position of the sample surface are obtained in the particular x-y-position, by identifying the sharpest image in each image stack. The sharpest image is selected by analyzing the contrast of the edges in each image of the image stack. To exclude perturbation effects introduced by the camera, the image sharpness integrated over the whole image is taken into account.

For each image the matrix $\mathbf{A}$ contains the gray-scale values in each pixel. The edge strength and thus the image sharpness can be calculating by first-order derivative differential edge detection performed by convolving (symbol $*$)

[1] Although the Photonic Professional® system provides an inbuilt autofocusing element which could be used for tilt-correction on a planar unstructured substrate, this element unfortunately fails for structured surfaces such as SOI-chips or laser dies.

each image matrix $\mathbf{A}$ with the Sobel gradient operators $\mathbf{S}_x$ and $\mathbf{S}_y$ [42], [43] resulting in two matrices $\mathbf{G}_x$ and $\mathbf{G}_y$ which contain the horizontal and vertical gradients in each pixel,

$$\mathbf{G}_x = \mathbf{S}_x * \mathbf{A}, \tag{2.1}$$

$$\mathbf{G}_y = \mathbf{S}_y * \mathbf{A}. \tag{2.2}$$

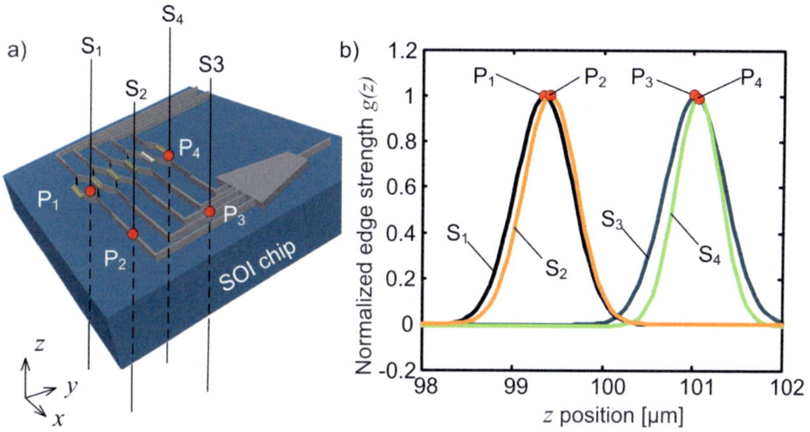

Fig. 2.1: Tilt correction (a) In four different x-y-positions of the piezo system images of the sample surface are taken at different z-heights. The sharpest image in each x-y-position is obtained directly at the sample surface. (b) Normalized fit of the edge strength g as a function of z-height for a tilted SOI chip. The sample surface z-position can be identified as the maximum of $g(z)$. The sample surface is remodeled by a plane fitted through the four resulting points $P'_1 ... P'_4 = (x_1, y_1, z_1)...(x_4, y_4, z_4)$ on the sample surface.

Using the Sobel operators, the gradient of the pixel intensity is calculated across three neighboring lines or columns where the center pixels have a weight of zero and are thus neglected. The Sobel operators $\mathbf{S}_x$ and $\mathbf{S}_y$ are given by:

$$\mathbf{S}_x = \begin{pmatrix} -1 & 0 & 1 \\ -2 & 0 & 2 \\ -1 & 0 & 1 \end{pmatrix} \tag{2.3}$$

$$\mathbf{S}_y = \begin{pmatrix} -1 & -2 & -1 \\ 0 & 0 & 0 \\ 1 & 2 & 1 \end{pmatrix} \tag{2.4}$$

$$g = \sum_{\mathrm{i,j}} \sqrt{\mathbf{G}_{x,\mathrm{ij}}^2 + \mathbf{G}_{y,\mathrm{ij}}^2} \tag{2.5}$$

The overall edge strength g for each image is obtained by pixel-wise summing the root mean square of the gradient matrices $\mathbf{G}_x$ and $\mathbf{G}_y$. A Gaussian fit of the edge strength $g(z)$ *vs.* z-height in every image stack is used to identify the sample surface positions $z_1 ... z_4$. The sample surface coincides with the sharpest image and thus the maximum of the Gaussian fit. Sample tilt is obtained in a following step by fitting a plane through the measured reference points $P'_1 ... P'_4 = (x_1, y_1, z_1) ... (x_4, y_4, z_4)$. In Fig. 2.1. b the analyzed edge strength, versus z-position is plotted for four points $P_1 ... P_4$ on the sample surface of a tilted SOI chip. With the knowledge of the sample orientation with respect to the optical axis of the lithography setup, the z-position of the chip surface in each x-y-position in the writing field can now be calculated.

2. Coordinate Transformation:

In the second step of the port detection algorithm a coordinate transformation is performed that correlates the position of an object in the camera image to its position expressed in lithography machine coordinates. All coordinate systems involved are assumed to be right-handed orthonormal systems.

The lithography coordinates of a specific point $\vec{r}_L = (x_L, y_L, z_L)^{\mathrm{T}}$ on a sample surface correspond to the position the piezo transducers need to address in order to move the specific point into the laser focus and thus expose it during the writing step. The axes of the lithography coordinates lie parallel to the piezo-transducer axis, with the origin coinciding with the piezo's zero position. The height z_L at which the polymerization takes place is dictated by the position of the laser focus with respect to the sample surface.

The axes of the camera coordinate system are given by the projection of the camera chip axis to the plane of sharpest sight $z_L + z_{LC}$. For a well-aligned lithography system the offset between lithography height and the sharpest image is $z_{LC} = 0$. As already mentioned in the previous section, the *z*-axis of the camera coordinate system is assumed to be parallel to the piezo *z*-axis and thus parallel to the lithography *z*-axis. The quantities IDx, IDy denote the discrete pixel numbers; *a* and *b* are physical dimensions of a single pixel. Further, a vector $\vec{r}_C$ in camera coordinates is defined by:

$$\vec{r}_C = \begin{pmatrix} a \cdot \mathrm{IDx} \\ b \cdot \mathrm{IDy} \\ z_L + z_{LC} \end{pmatrix} \tag{2.6}$$

To perform a coordinate transformation between the lithography coordinate system and the camera coordinate system we first need to identify the offset between the origins of both coordinate systems $\vec{r}_{c,0}$. To obtain $\vec{r}_{c,0}$, an image of the laser spot in its rest position is taken, i. e. the position in which the laser rests and no axis is moved. The centroid of this very bright laser spot corresponds to the offset vector $\vec{r}_{c,0}$ of the lithography coordinate system expressed in camera coordinates. The transformation $\mathbf{T}_{CL}$ between camera and lithography coordinates and the backward transformation $\mathbf{T}_{LC}$ can then be given in the form:

$$\begin{pmatrix} x_L \\ y_L \\ z_L \end{pmatrix} = \vec{r}_L \underset{\mathbf{T}_{CL}}{\overset{\mathbf{T}_{LC}}{\rightleftarrows}} \vec{r}_C = \begin{pmatrix} x_C = a \cdot \mathrm{IDx} \\ y_C = b \cdot \mathrm{IDy} \\ z_L + z_{LC} \end{pmatrix} \tag{2.7}$$

$$\vec{r}_L = \mathbf{M}_{LC} \cdot \vec{r}_c + \vec{r}_{c,0}. \tag{2.8}$$

Since both coordinate systems are parallel in their z-axis, the rotation matrix $\mathbf{M}_{LC}$ is given by:

$$\mathbf{M}_{LC} = \begin{pmatrix} \cos\theta & -\sin\theta & 0 \\ \sin\theta & \cos\theta & 0 \\ 0 & 0 & 1 \end{pmatrix} \tag{2.9}$$

The rotation matrix $\mathbf{M}_{LC}$ with the angle θ between the lithography coordinate system and the camera coordinate system can be obtained by taking two images of a single specific feature at a defined translation $\Delta\vec{r}_L$ in the lithography coordinate system. The two images are stitched together such that the feature coincides. This can be either performed manually or by automatic image stitching, exploiting correlation functions. From the resulting stitched image the image shift $\Delta\vec{r}_C = (\Delta x_C, \Delta y_C, 0)$ in camera coordinates is obtained.

The rotation matrix $\mathbf{M}_{LC}$ can be calculated from:

$$\vec{r}_L + \Delta\vec{r}_L = \mathbf{M}_{LC} \cdot (\vec{r}_C + \Delta\vec{r}_C) + \vec{r}_{LC}. \tag{2.10}$$

With Eq. (2.8) one obtains:

$$\Delta\vec{r}_L = \mathbf{M}_{LC} \cdot \Delta\vec{r}_C. \tag{2.11}$$

For a translation $\Delta\vec{r}_L$ along the lithography x-axis, the angle θ can be calculated, as it is displayed in Fig. 2.2:

$$\tan(\theta) = \frac{\Delta y_c}{\Delta x_c}. \tag{2.12}$$

After having obtained the transformation between camera and lithography coordinate system, several images can now be assembled to generate an overview of the whole writing field. This becomes necessary since the overall writing field of the conventional Photonic Professional® system measures 300 µm × 300 µm × 300 µm whereas the viewing field of the camera only measures 130 µm×100 µm. In the employed algorithm, 36 x-y-positions $P_1 ... P_{36}$ in the lithography coordinate system are chosen on a 50 µm grid. In these positions images are taken and used to generate the overview image.

Taking into account the sample tilt evaluated in the tilt correction step, the corresponding z-height for each of these positions is calculated. In the 36 resulting points $P_1'...P'_{36}$ images are taken. The single images are combined to an overview image, as can be seen in Fig. 2.3. Having identified the tilt of a sample with respect to the piezo z-axis in the tilt correction step, and having found a transformation between camera and lithography coordinate system, the lithography coordinates (x_L, y_L, z_L) of each point on the sample can now be calculated.

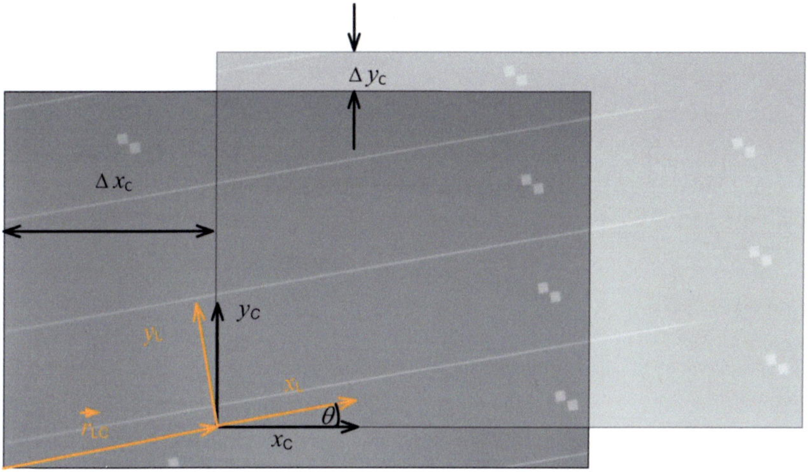

Fig. 2.2: Stitching of two single images with a fixed translation $\vec{r}_{LC}$. Here $\vec{r}_{LC}$ is chosen parallel to the lithography x-axis. The lithography coordinate system and the camera coordinate system are rotated by a specific angle θ with respect to each other.

3. Marker-based port detection:

As the last step, the algorithm used to analyze the assembled image for marker structures is discussed. In our configuration, the markers under investigation consist of two silicon squares. Those squares are fabricated in the same etch step as the SOI waveguides and are therefore located in the same z-layer. The squares have a side-length of 2.5 µm × 2.5 µm and touch in one corner along their diagonal.

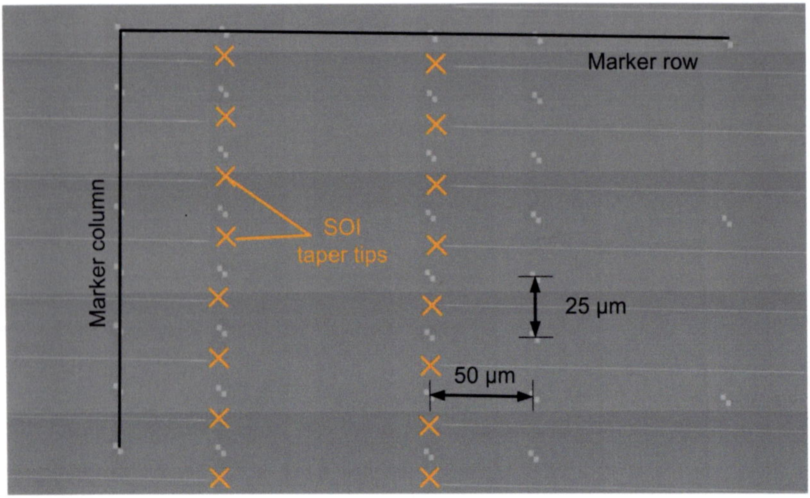

Fig. 2.3: Overview of the whole writing field by stitching of an array of single images. This image can now be used for marker based port detection. Markers are arranged in columns at a distance of 25 µm. Each waveguide tip lies in the middle between two neighboring markers. Rows of neighboring markers at a distance of 50 µm can be used to identify the waveguide direction.

To detect the markers, first, the assembled grayscale image is converted to a black/white image. The conversion threshold has to be adapted depending on the lighting conditions. Since silicon parts provide a higher reflectivity than the surrounding material, they appear white in the black/white image. In a next step, all connected white areas are identified. The areas are evaluated by their bounding box size, i. e. by the smallest rectangle containing the entire area and by their filling factor, i. e. by the ratio between white and black pixels inside the bounding box. Since every marker consists of two silicon squares, single squares are detected separately. It is decided whether or not a marker is present from a white area with a bounding box size of 20-50 pixels in horizontal and vertical direction and a filling factor larger than 35 %. These conditions are selected in order to be able to detect straight as well as tilted markers. Fig. 2.4 visualizes the marker together with its bounding box and center of mass. The presented values apply for an objective with 100× magnification. For objectives with a different magnification, the values have to be adapted.

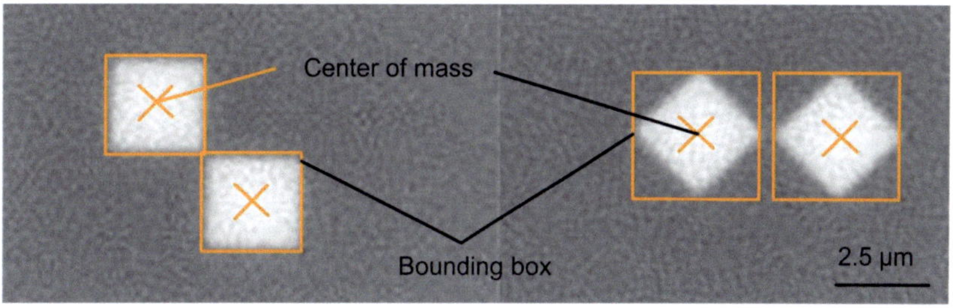

Fig. 2.4: Bounding boxes and centers of mass for a marker oriented parallel to the camera coordinate system and for a tilted marker under 45°.

Having identified the single squares of a marker, the center of mass for all white areas fulfilling the above-mentioned criteria is calculated. If the distance between the center of mass for two neighboring areas is smaller than 40 pixels the two areas are identified as the two connected silicon squares of one marker. The marker position is then defined as the middle between the center of mass of the two squares. These pixel positions are then transferred into lithography coordinates by Eq. (2.8).

Since the layout of the chip and the relative positioning of markers and waveguide outlets are precisely known, the waveguide positions and directions can now be directly calculated from the detected marker positions. For the SOI chip layout predominantly used in this work, the detected markers are sorted in columns by finding neighboring markers at a distance of 25 µm, see Fig. 2.3. Each waveguide tip lies in the middle between two neighboring markers. Rows of neighboring markers at a distance of 50 µm are used to identify the waveguide direction. The knowledge of the waveguide positions and directions is in the following used to calculate a matching PWB trajectory.

2.2 Trajectory

For minimum optical loss, the trajectory of a PWB axis has to obey a number of constraints. To begin with, the end-faces of the PWB must overlap with the connecting waveguides on both sides. Next, the starting and ending orientation of the PWB axes have to coincide with the connecting waveguide axes. Further, the trajectory of the PWB axis is to be chosen such that intersections with obstacles like fibers, chip edges or other PWB are avoided. Finally, increased losses by a strong curvature of the trajectory and a large length should be avoided, and a suitable compromise between structure length and bend radius has to be found.

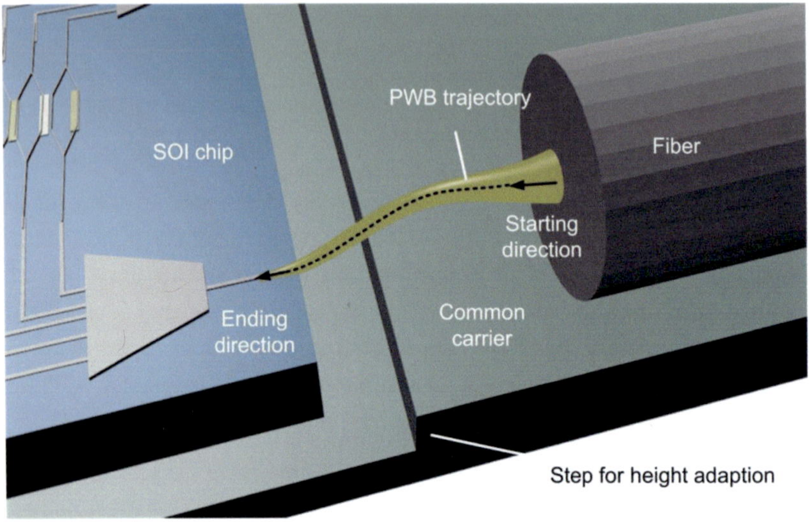

Fig. 2.5: Schematic of a photonic wire bond between fiber and silicon-on-insulator (SOI) chip [J1]. The PWB trajectory is selected, such that it adapts to the positions and the optical axes of the structures to be connected. Obstacles like chip or fiber edges must be avoided, and a good compromise is to be found between long interconnects and sharp waveguide bends.

To fulfill the requirements listed above, an optimization process has to be employed. In order to be able to minimize the optical transmission loss, an appropriate loss model for PWB trajectories is required. While the theory of in-plane waveguide bends is well developed [44]–[47], no theory is available for waveguide axes that are bent in three dimensions. Work is in progress to find the parameters of a closed-form empirical function for the photonic wire bond trajectory related losses. As a first attempt, a trajectory planning algorithm has been developed that takes into account bending as well as length-dependent losses, and thus finds a suitable compromise between long structures and sharp bends.

2.2.1 Loss Mechanisms

Theoretically, optical losses in an ideal, straight waveguide should be zero. However, even the transmission through a straight waveguide leads to non-negligible losses.

These losses are caused by material absorption and scattering loss. Material absorption is an inherent loss mechanism in all waveguides. Whereas glass is an almost ideal material for the communication at 1550 nm wavelength, most polymer materials cause more pronounced material losses due to the excitation of molecule oscillations that are present at wavelengths typically used for optical communications [48], [49]. For length-dependent losses, the optical power P along the light propagation direction x can be described by an exponential decay of the initial power P_0. The quantity $P(x)$ denotes the power at position x flowing in x-direction, and α_{mat} is the power attenuation coefficient representing the power loss per unit length,

$$P(x) = P_0 \, \mathrm{e}^{-\alpha_{\text{mat}} x} . \tag{2.13}$$

Material absorption amounts to typical values of 2-3 dB/cm [50] for epoxies like SU-8. The loss contribution by material absorption in the case of photonic wire bonding however, is very small since the distance a photonic wire bond spans lies in the range of a few hundred micrometers. This corresponds to a loss contribution of 0.02 - 0.05 dB per PWB and can be tolerated.

Adding up to the material-dependent absorption losses, Rayleigh scattering and surface scattering are additional length dependent loss contributions. Rayleigh scattering can be caused by inhomogeneities in the waveguide that are small compared to the material wavelength, such as small refractive index differences that might be caused by inhomogeneous crosslinking in the core of the photonic wire bond or smallest dirt particles in the photo resist material. In order to avoid Rayleigh scattering, a homogeneous crosslinking or polymerization as well as clean samples and chemicals are important. The exact loss contribution of Rayleigh scattering in photonic wire bonds, however, is not clear at the present stage. Surface scattering occurs at surface roughness or inhomogeneities in the shell of the photonic wire bond. It is therefore directly linked to the surface quality of the PWB and thus to the writing strategies discussed in Section 2.4. Again, at this moment the exact contributions to loss arising from these two mechanisms are not yet quantified, since the amount and the reproducibility of available measurement data are insufficient.

Another source of optical loss that is related to the PWB trajectory is bending loss. To explain the source of bending loss, it is important to understand that in bent waveguides leaky modes evolve instead of guided waveguide modes. Let us assume an optical waveguide with a constant curvature and a fixed waveguide width w. The evanescent part of the waveguide mode extends beyond the waveguide boundary. In this waveguide, the electrical field components on the outside of the bent waveguide travel a longer distance compared to the field components inside the core. In order to stay in phase with the field components inside the waveguide core and thus maintain a planar wave front, the outer field components would have to travel at higher velocities. Furthermore, the velocity of these field components would even have to increase with the distance from the core. Since the velocity in the outside region cannot exceed the local speed of light, the equi-phase front starts to bend, and energy is transferred to radiative modes, see Fig. 2.6.

Bending loss strongly depends on the curvature $\kappa = 1/R_c$, where R_c is the curvature radius, on the dimensions of the waveguide, and on the refractive index n_{co} of the core and n_{cl} of the cladding. All of these factors influence the confinement of the optical mode. For a higher refractive index contrast $\Delta = \left(n_1^2 - n_2^2\right)/2n_1^2$ the mode is more confined to the waveguide core, and smaller bend radii become possible. A detailed analysis of losses in optical waveguides is given by Marcatili *et al.* [46] who analyze in-plane bends of rectangular waveguides.

In general the trajectory $\vec{r}$ of a PWB does not lie in one plane, but follows a free-form curve in space. This means that the trajectory also includes torsion τ as well as transitions between waveguide sections of different curvature. While Marcatili *et al.* [46] give a useful description of in-plane curves and transitions, their work is not directly applicable to describe transitions in three-dimensional free form waveguides such as photonic wire bonds. Work is in progress to find a closed-form model to describe the losses of three-dimensional free-form waveguides, however, this model is not found yet and the search for it exceeds the framework of this thesis.

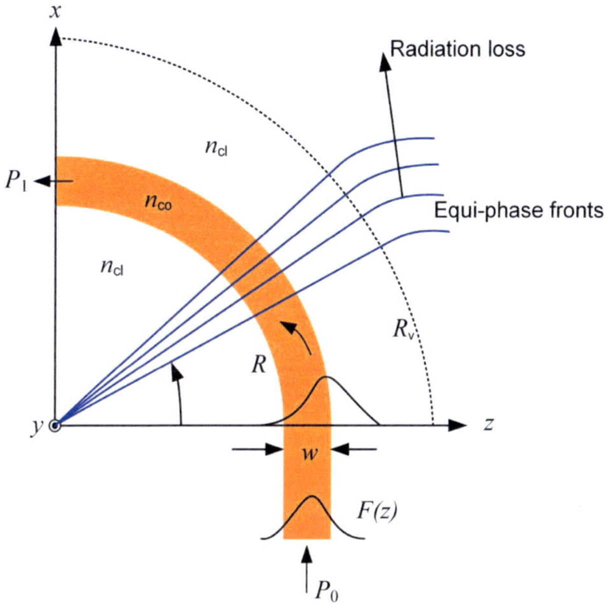

Fig. 2.6: Waveguide with constant curvature adapted from [46]. The waveguide core has a refractive index n_{co} and is embedded in a material with lower refractive index n_{cl}. Due to the waveguide bend the optical field distribution $F(z)$ guided inside the waveguide is shifted outwards. Significant field components lie outside of the waveguide cross-section w. Starting at a radius R_v the equiphase fronts start to bend and energy is transferred to radiative modes, causing the optical output power P_1 to be smaller than the optical input power P_0.

In addition to that lack of a theoretical model, the individual loss contributions that sum up to the overall PWB losses cannot be distinguished experimentally at the present stage. This is due to the fact that photonic wire bonds cannot yet be fabricated with the required reproducibility and that the PWB losses are at the moment dominated by the interface losses of the PWB towards the optical feed waveguides.

2.2.2 Trajectory Planning Algorithm

In the framework of this thesis, however, we developed a first trajectory planning algorithm that tries to balance polarization-independent bending losses as well as length depending losses, such as material absorption and surface scattering losses. So far, the algorithm does not take into account transition losses and torsion losses nor polarization-dependent bend losses and can therefore not describe the trajectory related losses of a three-dimensional free form waveguide completely. At the present stage it is not yet clear, if a closed-form analytical model for the losses of a 3D freeform waveguide solely based on the trajectory can be found. However, models that also take into account polarization, torsion as well as transitions are currently investigated in the framework of a large-scale research collaboration.

In analogy to [51], our simplified loss model describes the power $P(l)$ at the end of the trajectory $\vec{r}(s)$ with the length l by an exponential decay. The vector $\vec{r}(s)$ is given in Cartesian coordinates and $P(0)$ denotes the input power at the trajectory's start. Material absorption and surface scattering losses are described by α, and the variable s describes the arc length of the trajectory. The position vector along the trajectory is given by $\vec{r}$. The free parameters γ and q are adjustment factors that balance curvature dependent losses vs. length dependent losses and can be used as fitting parameters to adapt the model to simulated or measured value. The power $P(l)$ can be written as

$$P(l)=P(0)\exp\left[-\int_{0}^{l}\left(\alpha+\gamma\kappa(s)^{q}\right)\cdot\mathrm{d}s\right]. \tag{2.14}$$

The waveguide curvature $\kappa(s)$ can be calculated by the second derivative of $\vec{r}$ with respect to the arc length and represents the inverse of the radius of curvature R_c,

$$\kappa(s)=\frac{1}{R_c}=\left|\frac{\mathrm{d}^2\vec{r}(s)}{\mathrm{d}s^2}\right|. \tag{2.15}$$

With this Eq. (2.14) can be rewritten into the form

$$P(l)=P(0)\exp\left[-\int_0^l\left(\alpha+\gamma\left|\frac{\mathrm{d}^2\vec{r}(s)}{\mathrm{d}s^2}\right|^q\right)\cdot\mathrm{d}s\right]. \tag{2.16}$$

To find an optimum low loss trajectory, we use a calculus of variations of the functional $J(\vec{r})$ defined in Eq.(2.17); maximum power transmission is expected at minimum values of $J(\vec{r})$.

$$J(\vec{r})=\int_0^l\left(\alpha+\gamma\cdot\kappa(s)^q\right)\mathrm{d}s, \tag{2.17}$$

Since the length of the trajectory l is one of the parameters that varies during the optimization process, a re-parameterization of the loss model is required, such that the loss functional $J(\vec{r})$ becomes independent of the length l. This can be obtained by writing the arc length s in dependency of a curve parameter t that ranges from 0 to 1. Here the vectors $\dot{\vec{r}},\ddot{\vec{r}},\dddot{\vec{r}}$ represent the first, second and third derivative of the trajectory with respect to the curve parameter t.

$$\mathrm{d}s=\left|\dot{\vec{r}}\right|\mathrm{d}t. \tag{2.18}$$

According to [52] the curvature as a function of the curve parameter t can be written as:

$$\kappa(t)=\frac{\left|\dot{\vec{r}}(t)\times\ddot{\vec{r}}(t)\right|}{\left|\dot{\vec{r}}(t)\right|^3}. \tag{2.19}$$

Using this re-parameterization the functional $J(\vec{r})$ is obtained:

$$J(\vec{r})=\int_0^1\left(\left|\dot{\vec{r}}(t)\right|\cdot\alpha+\left|\dot{\vec{r}}(t)\right|\cdot\gamma\left(\frac{\left|\dot{\vec{r}}(t)\times\ddot{\vec{r}}(t)\right|}{\left|\dot{\vec{r}}(t)\right|^3}\right)^q\right)\mathrm{d}t. \tag{2.20}$$

In the given procedure the position vector $\vec{r}$ is defined to be

$$\vec{r}\left(x(t),y(t),z(t)\right)=\left(x(t),y(t),z(t)\right)^{\mathrm{T}}. \tag{2.21}$$

In our approach, the functions *x, y,* and *z* are selected to be fifth order polynomials. Here the expression for the *x*-component is given as an example, the *y*- and *z*- component are similarly defined,

$$x(t)=a_{x0}t^5+a_{x1}t^4+a_{x2}t^3+a_{x3}t^2+a_{x4}t+a_{x5}. \tag{2.22}$$

The discussed prototype algorithm aims at a trajectory adapted to the positions, as well as to the starting and ending directions of the interconnect waveguides. These positions and directions are measured as described in Section 2.1. The vector $\vec{r}_0$ is the position of the first interconnect waveguide, and $\vec{r}_1$ is the position of the second interconnect waveguide. The waveguides starting and ending directions are given by the unity vectors $\vec{v}_0$ and $\vec{v}_1$. The trajectory is adapted to the positions and directions of the feed-waveguides if the following equa-tions hold

$$\begin{aligned}
&\vec{r}(t=0)=\vec{r}_0,\\
&\frac{\dot{\vec{r}}(t=0)}{\left|\dot{\vec{r}}(t=0)\right|}=\vec{v}_0,\\
&\vec{r}(t=1)=\vec{r}_1,\\
&\frac{\dot{\vec{r}}(t=1)}{\left|\dot{\vec{r}}(t=1)\right|}=\vec{v}_1,
\end{aligned} \tag{2.23}$$

Since the adaption of the polynomial to the boundary condition only requires four degrees of freedom in each direction, there is still an infinite number of fifth order polynomials that fulfill these requirements. These additional degrees of freedom are used to minimize the functional $J(\vec{r})$ by a Broyden-Fletcher-Goldfarb-Shanno (BFGS) quasi-Newton algorithm [53] as implemented in the Matlab function *fminunc* [54],[55]. This optimizer applies a cubic line search procedure to find the minimum of a general function in several iteration steps. As a result of the optimization step, a PWB trajectory $\vec{r}(t)$ following a fifth order polynomial, avoiding both a long trajectory as well as a strong curvature is obtained in a case where no further constraints have to be met.

Trajectory calculation with obstacles

Our concept of photonic wire bonding has to provide the design freedom, to find suitable PWB trajectories also in the cases where the preferred unconstrained trajectory is blocked by obstacles. Those obstacles can be chip edges, other photonic wire bonds, electrical wires, or residuals of glue that has been used to mount the photonic components on the common carrier as described in Chapter 1. An intersection of the trajectory with these obstacles has to be avoided. This can be reached, by modifying the functional $J(\vec{r})$ from Eq.(2.20) by adding an additional factor $g(\vec{r})$:

$$J(\vec{r})=\int_0^1\left(|\dot{\vec{r}}(t)|\alpha+|\dot{\vec{r}}(t)|\gamma\left(\frac{|\dot{\vec{r}}(t)\times\ddot{\vec{r}}(t)|}{|\dot{\vec{r}}(t)|^3}\right)^q+g(\vec{r})\right)\mathrm{d}t \qquad (2.24)$$

$$g(\vec{r})=g_x(x,y)g_y(x,y)g(z). \qquad (2.25)$$

This factor $g(\vec{r})>0$ acts as a "penalty" function that artificially increases the value of $J(\vec{r})$ if the trajectory intersects with a "forbidden" area. It is important to mention that the penalty factor has to allow the chosen optimization algorithm to converge out of the undesired area. This is not possible using a constant penalty function. In the presented approach $g(\vec{r})$ is chosen to be a continuous function with its highest value in the inner core of the "forbidden" area and decreasing values towards the outside. Since the avoidance of chip edges is the most common practical case, the penalty function is chosen to cover a prism shaped obstacle, defined by four corner points that describe the upper surface of the obstacle. These four corner points are specified by a user interaction. The penalty function $g(\vec{r})=g_x(x,y)g_y(x,y)g(z)$ is given by

$$g_x(x,y)=\left(\frac{1}{\sigma\sqrt{2\pi}}\right)^2\int_{-\infty}^{x}e^{\frac{-(\tau-x_{p0}(x,y))^2}{2\sigma^2}}\mathrm{d}\tau\int_{-\infty}^{x}e^{\frac{-(-\tau+x_{p1}(x,y))^2}{2\sigma^2}}\mathrm{d}\tau,$$

$$g_y(x,y)=\left(\frac{1}{\sigma\sqrt{2\pi}}\right)^2\int_{-\infty}^{y}e^{\frac{-(\tau-y_{p0}(x,y))^2}{2\sigma^2}}\mathrm{d}\tau\int_{-\infty}^{y}e^{\frac{-(-\tau+y_{p1}(x,y))^2}{2\sigma^2}}\mathrm{d}\tau, \quad (2.25)$$

$$g(z,z_{p0})=\exp\left(1-\frac{z}{z_{p0}}\right).$$

The parameters $x_{p0},y_{p0},x_{p1},y_{p1}$ are given by the intersection between the edges of the prism containing the forbidden area with lines through the point $B=(x,y)^{\mathrm{T}}$ that lie parallel to the x-axes, and y-axis respectively as can be seen in Fig. 2.7.

$$\begin{aligned}\vec{r}_{p0}(x,y)&=\left(x_{p0}(x,y),y_{p0}(x,y)\right)^{\mathrm{T}},\\ \vec{r}_{p1}(x,y)&=\left(x_{p1}(x,y),y_{p1}(x,y)\right)^{\mathrm{T}}\end{aligned} \quad (2.26)$$

The slope of the function is defined by the parameter σ that can also be understood as the standard deviation of the cumulative distribution function. Fig. **2.7** gives an image of the penalty function $g(x,y)$ as well as $g(z)$.

Eq. (2.24) can be used for numerical optimization in the same way as it has been described for the case of a trajectory without obstacles.

As a result of the discussed algorithm the PWB trajectory is again obtained in the form of a fifth order polynomial in three dimensions, parameterized in terms of a curve parameter t. The trajectory $r(t)$ starts and ends at the waveguide interconnect positions dictated by the feed waveguides and features the same waveguide direction in these points. The resulting polynomial represents a suitable compromise between a long trajectory and sharp waveguide bends. Based on this trajectory a 3D PWB model is generated as it is discussed in Section 2.3.

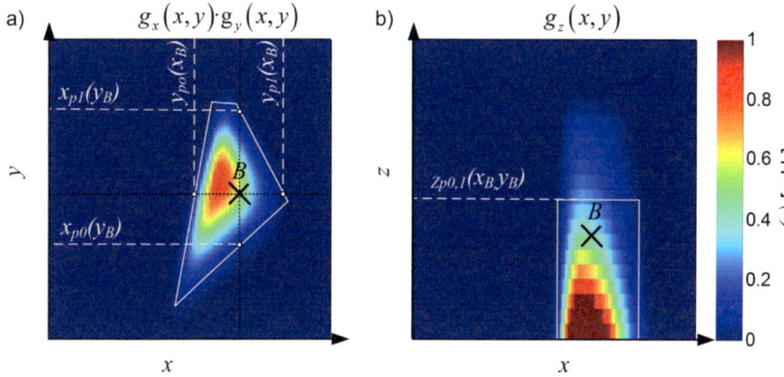

Fig. 2.7: Normalized penalty function $g(\vec{r})$ for a prism shaped obstacle in *x*-*y*-cut (a) and *z*-*x*-cut (b) The boundary of the obstacle is indicated by the white solid line.

2.3 Generation of the Three-Dimensional Photonic Wire Bond Shape

The last section has given an insight into the three-dimensional trajectory planning algorithm. The trajectory is the core of the photonic wire bond and based on it, the three-dimensional shape of the photonic wire bond is generated. This section discusses the generation of the 3D PWB shape. The calculation approach works in several steps. It allows varying the outer shape of the PWB waveguide and especially the waveguide cross-section along the trajectory.

This is of special importance for adapting the waveguide mode of the photonic wire bond to the mode of an interconnecting waveguide. Modefield adaption is mostly reached by tapered sections that are realized as an integral part of the PWB; both, up- and down-tapers[2] can be generated.

In general, the photonic wire bond can be divided into five segments: The start taper, the start conversion, the waveguide section, the end conversion and the end taper. Whereas the waveguide section of the PWB follows the pre-calculated trajectory, start and end taper follow straight lines with the direction given by the first derivative of the trajectory in the start and end point. The dimensions of the taper can be chosen by user interaction and based on the selected dimensions straight segments of specific length are then added to the trajectory. The segments where the outer shape of the PWB changes, for example from rectangular to circular, are denoted as conversion.

The calculation of the 3D photonic wire bond shape works in several steps:

First, the trajectory, as it is shown in Fig. 2.8, including the straight taper sections, is split into segments, by calculating L equidistant nodes $\vec{r}(t_l)=(x_l,y_l,z_l)^{\mathrm{T}}|_{l=1...L}$ on the trajectory. The number L can be freely chosen here. While smaller numbers of L lead to a faster calculation of the final PWB structure, it is beneficial to choose L, such that the distance between two neighboring nodes along the trajectory does not exceed 1 µm in order to smoothly reconstruct the trajectory.

---

2 An up-taper is a taper where the waveguide cross-section in the taper end, i.e., the end of the waveguide, is larger than the waveguide cross-section in the untapered part. This is the case, e. g., if a cylindrical waveguide with a waveguide diameter of 2 µm is tapered up to a waveguide diameter of 5 µm towards the waveguide end.
A down-taper is a taper where the taper tip is smaller than the waveguide cross-section in an un-tapered section. This is often the case in silicon on insulator waveguides. Here the waveguide width in the un-taperd case amounts to 500 nm, the taper tip however amounts to approximately 100 nm. In the case of silicon on insulator waveguides the height of the tapers and waveguide section is the same.

Second, the tangential vector of the trajectory in each node $\vec{r}(t_l)=(x_l,y_l,z_l)^{\mathrm{T}}$ is calculated by the first order derivative of the trajectory $\vec{v_l}=\mathrm{d}\vec{r}(t)/\mathrm{d}t\big|_{t=t_l}$, see Fig. 2.9.

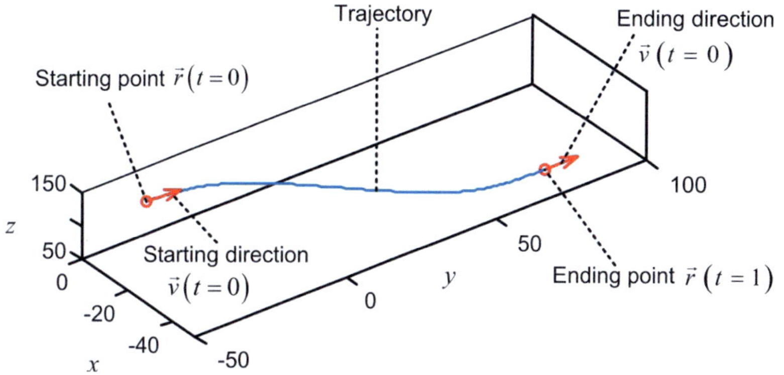

Fig. 2.8: Trajectory of a photonic wire bond between a starting point and an ending point: The trajectory starts at the starting point with the direction dictated by the input waveguide, and ends with the ending direction adapted to the output waveguide axis direction.

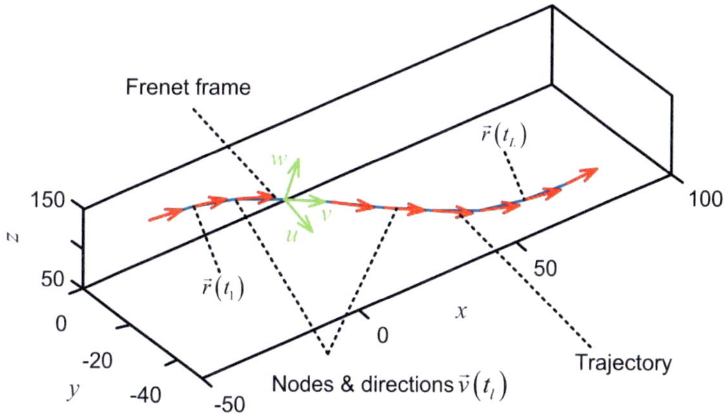

Fig. 2.9: The trajectory is divided into equidistant segments. In each of the nodes $\vec{r}(t_l)=(x_l,y_l,z_l)^{\mathrm{T}}$ the trajectories' direction $\vec{v}(t_l)$ is calculated by the first derivative of the trajectory in three dimensions.

In the third step, two-dimensional contours are calculated to be perpendicular to the tangential vector of the trajectory on every node. These contours specify the outer shape of the photonic wire bond as can be seen in Fig. 2.13. Each outer contour consists of *N* points, where $N=4n$ with $n\in\mathbb{N}$. For a rectangular contour each side of the rectangle is attributed $N/4$ points. For an elliptic shape the *N* points are angle-wise equally distributed. If required, morphing is performed in this step.

In some applications it is beneficial to use PWB tapers with rectangular cross section, while in others an elliptical taper shape is required. To convert a rectangular cross-section to an elliptic one as well as to change the cross-section area, a number of *M* successive layers is required. The starting and ending shape of the conversion are first calculated as a list of points in the *uw*-plane od the local coordinate system *uwv*, with the center of gravity of both contours coinciding with the origin of the coordinate system, see Fig. 2.10. a.

While the center of gravity of the starting contour remains positioned at $(0,0,0)$, the center of gravity of the ending contour is shifted to $(0,0,M)$. For each of the *N* points of the first contour a straight connecting line is calculated between the first and last contour, see Fig. 2.10. b.

These *N* lines are then intersected at $M-1$ different heights, resulting in $M-1$ new contours defined by *N* contour points as can be seen in Fig. 2.11. After this operation the contour points of every layer are projected back into the *uw*-plane as is displayed in Fig. 2.12.

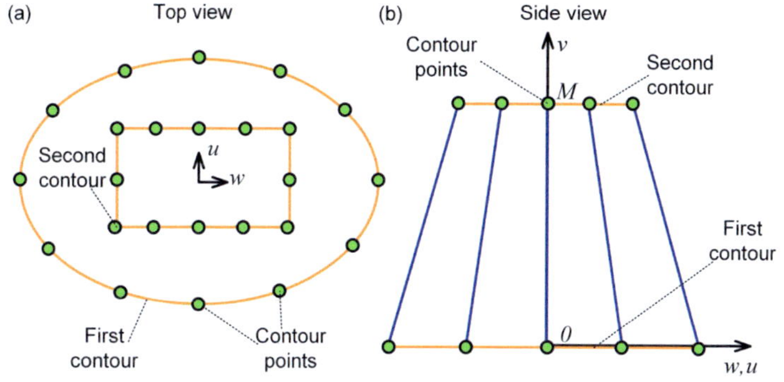

Fig. 2.10: Schematic drawing of the morphing between a first elliptical and a second rectangular contour; Step 1: For the morphing within M layers the z position of the first contour layer is chosen to be $v = 0$ whereas the last layer is positioned at $w = 0$, $u = 0, v = M$. For each of the N points of the first contour a connecting line between the points in the first and last contour is calculated.

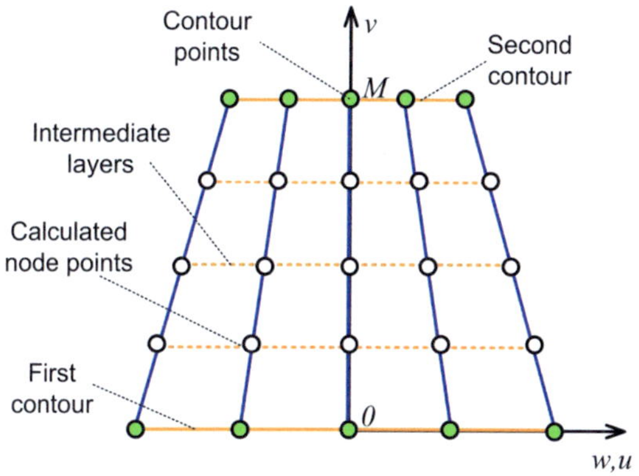

Fig. 2.11: Schematic drawing of the morphing between a first elliptical and a second rectangular contour; Step 2: The interconnect lines generated between the first and last contour layer are intersected at $M-1$ intermediate layers at equidistant heights. Each intersection between the height layer and the interconnect lines is defined as a new node point. For each of the intermediate layers a new contour defined by N contour points results.

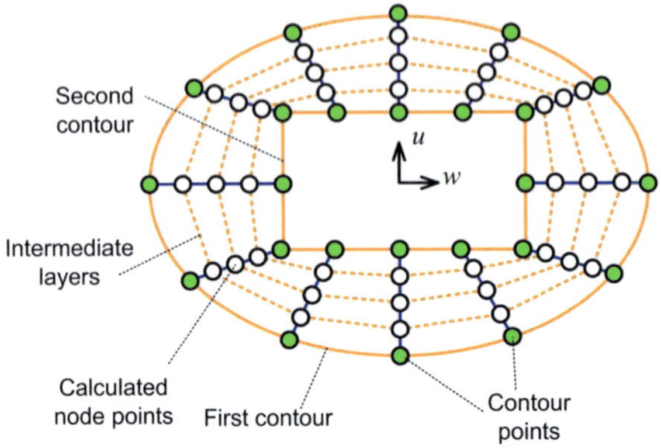

Fig. 2.12: Schematic drawing of the morphing between a first elliptical and a second rectangular contour; Step 3: After the calculation of the intermediate layers new resulting contour points of every layer are projected back into the *uw*-plane.

After having calculated all the planar contours in the *uw*-plane, the single contours are rotated such that the normal vector of the contours coincides with the tangential vector of the trajectory at the appropriate node. In a next step the center of gravity of each contour is shifted to the appropriate node in *xyz*-coordinates as can be seen in Fig. 2.13.

After having calculated the outermost contour of the 3D PWB structure, the list of contour points is further processed depending on the chosen writing strategies that will be discussed in Section 2.4.

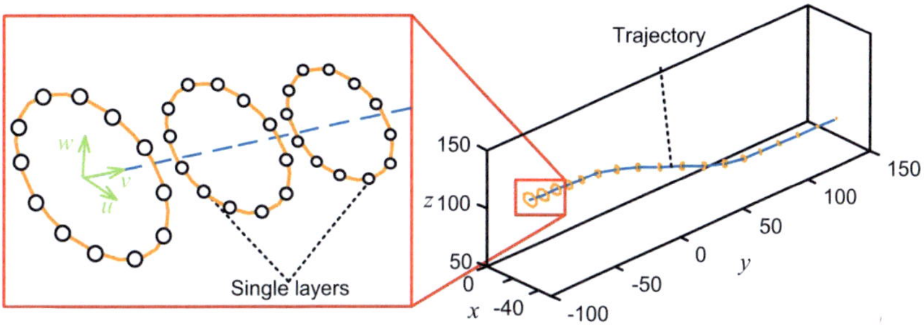

Fig. 2.13: Distribution of layers along the trajectory. For every segment of the trajectory the center of gravity of each contour is shifted to the appropriate position. The normals of the contours are rotated, such that they coincide with the tangential vector of the trajectory in the specific point.

To prepare the final 3D structure for writing, the outermost contour is filled, such that a solid structure is generated. This can be performed by repeating the described process multiple times, reducing the initial dimensions by a specified distance Δ after every step. To define the writing lines, the single contour points are then connected *via* straight lines, forming the contour of the PWB structure, as it is displayed in Fig. 2.14. This process is applied for the trajectory writing approach in Section 2.4.1.

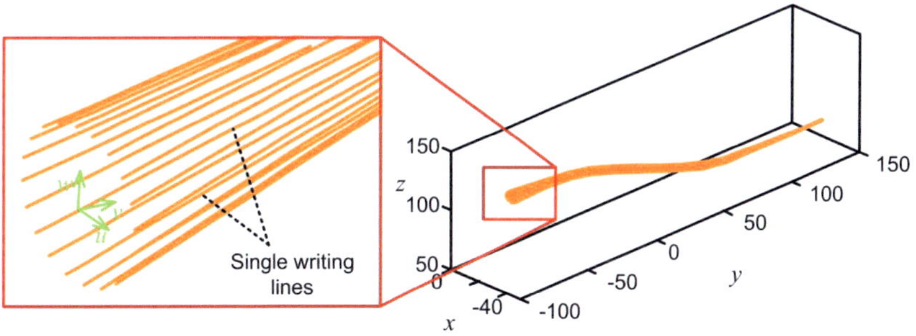

Fig. 2.14: Definition of single writing lines from the contour points calculated in Fig. 2.12. The single contour points are connected *via* straight lines, forming the outer contour of the PWB structure. Depending on the writing strategy chosen the list of contour points is further processed.

Surface Tesselation Language (stl) is a file format widely used in rapid prototyping systems. It describes the surface of a three-dimensional object by a list of vertices and their outward pointing normal vectors [117]. Structures in stl-format can be easily transferred to single writing lines parallel to the xy-plane of the lithography system. Our algorithm provides the option to export the 3D PWB shape to an stl-file. The export of the generated 3D PWB shape to stl-files is described in Appendix A.1. The 3D PWB shape can then be further processed by software tools such as DeScribe [56] to generate machine-readable data sets.

To ensure a secure adhesion of the photonic wire bond to the chip surface the described process provides the possibility to include anchor structures. These parts of the PWB are written into the substrate surface. Obviously, no polymerization can take place inside the substrate. This anchoring structure, however, ensures that even at a slight mismatch in the substrates measured and real z-position, an adhesion of the PWB to the substrate, e. g. an SOI chip, is guaranteed.

As an anchor structure at the SOI interface a prism shape is chosen, such that the anchor depth vanishes at the SOI taper tip and increases to the final chosen anchor depth h_{anchor} at the polymer taper tip. The anchor is deepest where the PWB structure is smallest as can be seen in Fig. 2.15. In the case of a fiber interface, the PWB taper is extended into the fiber with a uniform anchor depth.

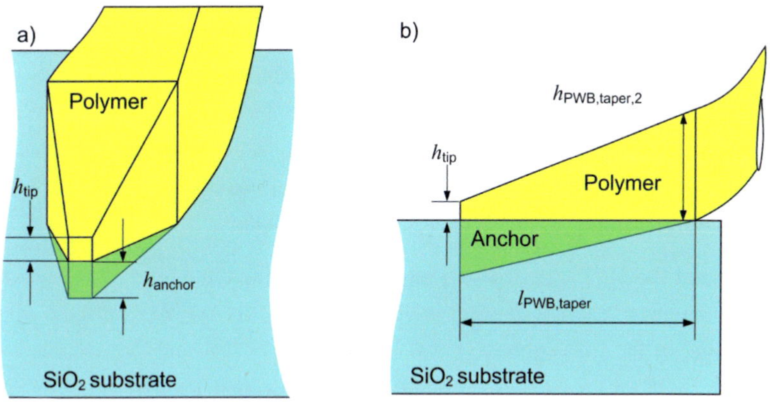

Fig. 2.15: Schematic illustration of the anchor structure beneath the polymer taper: a prism shape intersecting with the SiO_2 substrate is chosen. The depth h_{anchor} of the anchor varies between the taper tip and the taper end, such that the anchor depth is $h_{anchor} = 0$ at the SOI taper tip and increases to the final chosen final anchor depth at the polymer taper tip, thus the anchor is deepest where the taper cross-section is smallest.

2.4 Writing Strategies

The three-dimensional model of the photonic wire bond can now be used for structure generation, using the lithography system described in Section 1.3.3. The structure is fabricated from liquid or gel-like resist material, by exposing the resist at precisely defined locations in three dimensions as explained in Section 1.3.1. To this end, the lithography system performs a controlled movement of the systems focus through the volume of the resist material. The individual trace the focus follows is defined by the structure itself as well as by a specific writing strategy. The choice of an adequate writing strategy is depending on the underlying system, on the kinematics used and the properties of the resist material. In the following three different writing strategies are discussed.

2.4.1 Writing along the Trajectory

The most straightforward writing strategy to define the PWB structure is writing lines that lie more or less parallel to the center trajectory of the PWB, as can be seen in Fig. 2.16. This strategy, however, is only suitable for resist material that forms a solid or gel like matrix during sample preparation. An example for this is SU-8 that solidifies during pre-bake.

Following the 3D PWB structure algorithm described in Section 2.3 the export into a machine readable dataset is very easy. To this end the software selects one point of the first contour, i. e. N=const is selected, and then successively goes through all layers by incrementing L, connecting the Nth point of every layer with a straight line. Traveling through all contours, this results in a polygon chain in three dimensions that represents one writing line. The process is repeated for all points in the first contour. Since the 3D structure of the photonic wire bond consists of many nested substructures, the algorithm shrinks the outer contour and by a specified writing distance Δ and repeats the process until the distance between opposing sides of an ellipsoidal or rectangular cross-section fall below the specified writing distance.

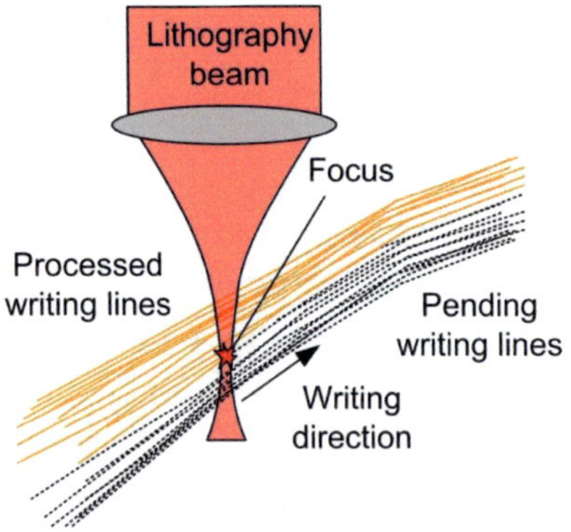

Fig. 2.16: Photonic wire bond generated from writing lines parallel to the center trajectory.

Before being exported to a machine readable dataset, neighboring lines that fall below the minimum distance, as well as single points are deleted. All remaining polygonal chains, each followed by a "write command", are then exported in the format required by the lithography system.

The great advantage of this writing strategy is the generation of a smooth waveguide surface. This is a consequence of the fact that all writing lines *a priori* lie parallel to the PWB surface. This advantage, however, only comes into effect when solid or gel-like resist materials like SU-8 [38] are chosen. The writing strategy is not suitable for liquid resists, due to the constant change of z-positions that even in the most recent modification of the lithography system are attributed to a movement of the sample with respect to the laser focus. In liquid resist this results in a fluid movement and the first voxel lines tend to drift apart as they are not attached to already polymerized material on their whole length. This results in undesired shapes that are not usable for optical waveguiding, as can be seen in Fig. 2.17.

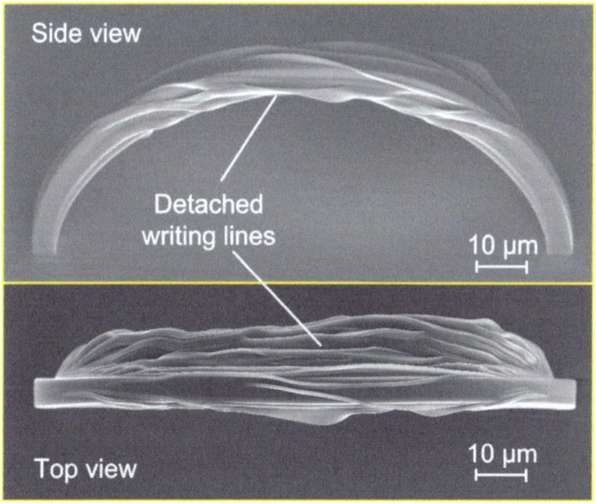

Fig. 2.17: Undesired movement of the resist during the lithography step. Due to the constant movement of the piezo systems a floating can occur inside a liquid resist that results in a detachment of single writing lines.

2.4.2 Helix Writing

When using resist materials which are viscous or liquid during the writing operation, a movement or floating of already exposed structural regions within the resist volume must be prevented.

One way to overcome this issue is helix writing, which is ideally suited for piezo-based lithography systems. The three-dimensional waveguide structure is defined from individual disks which are oriented perpendicular to the center line of the waveguide as shown in Fig. 2.18. The disks can be split up, for example, into straight segments or into concentric elliptical or circular spirals (helices). In the presented approach, the photonic wire bond is built up from helices that start alternating at the innermost, or at the outermost radius of the individual slice. The innermost point of every slice is located on the center trajectory and every slice is surrounded by a contour line with the designed circumference to ensure a smooth surface. With this method the PWB waveguide is fabricated starting with the connecting locations and is thus at all points in time fixedly connected to the substrates and/or the input and output waveguides.

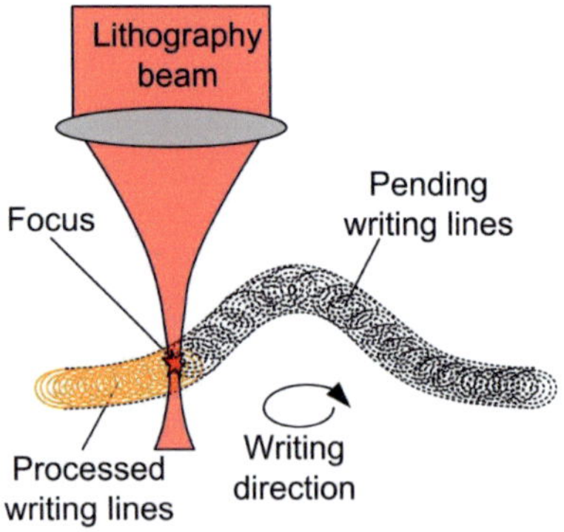

Fig. 2.18: Schematic of the helix writing strategy: The three-dimensional waveguide structure is defined from individual helices which are oriented perpendicular to the center line of the waveguide.

2.4.3 Line Writing

In line writing, the 3D model of the photonic wire bond is segmented into single lines, all being parallel to the *xy*- plane. In order to obtain those lines a 3D model file (e. g. an *.stl file) is first cut into single layers parallel to the *xy*-plane (slicing) with a distance of Δz along the *z*-axis. These layers are then further cut into individual lines (hatching) being parallel to either the *x*- or the *y*-axis with a hatching distance of Δz or Δy respectively. This technique is a common procedure in rapid prototyping systems and 3D printers. In the presented case the single writing lines are sorted, such that each layer consists of one writing line that fills the whole layer in a meander form.

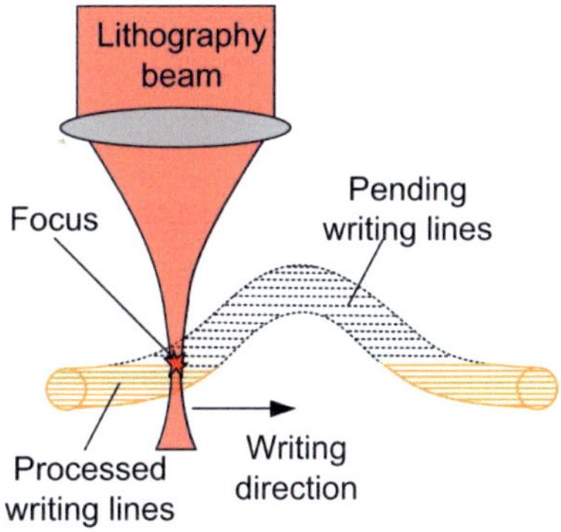

Fig. 2.19: Line writing strategy: Comparable to 3D printers in rapid prototyping

This strategy is ideally suited for galvanometer-mirror based lithography systems. As already explained in Section 1.3.3 these systems provide very high writing velocities in the *xy*-plane since they keep the sample in a fixed position and steer the laser beam with the help of fast moving mirror. The axial writing in these systems, however, still relies on the movement of the sample by piezo actuators. This movement is by factors slower than the *xy*-movement. To exploit the full velocity of these systems a constant movement of the *z*-axis has to be avoided and sliced writing as it is common in rapid prototyping systems lends itself to implementation.

One of the major drawbacks of slicing and hatching lies in enhanced surface roughness due to a discretization in the *z*-axis, see Fig. 2.20 left. This "staircase effect" can significantly increase optical losses, and especially appears in sections with a small slope. Several ways to avoid the staircase effect exist, the simplest one makes use of a very small discretization step Δz, which comes at the cost of increased writing time. More advanced techniques rely on the insertion of intermediate contour lines at sections with a shallow slope.

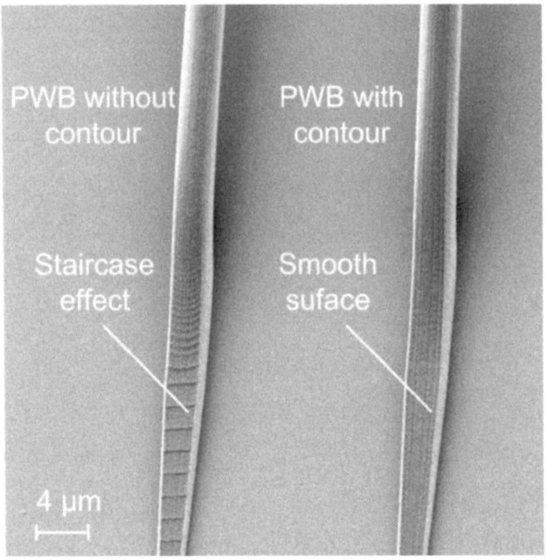

Fig. 2.20: Photonic wire bond with and without contour. Due to the height discretization in the slicing process the correct contour is not represented in the left PWB and staircase-like surfaces evolve especially in areas of the sample with a small slope as for example in the 3D double taper section. The PWB on the right-hand side is smoothed by adding an outer contour following the trajectory writing algorithm to the sliced inner structure. This technique requires the piezo transducer stages as well as the galvanometer-driven mirrors of the lithography system to be perfectly aligned.

Another potential way to smoothen the surface of a structure written in line writing technique is to add an outer contour following the trajectory writing algorithm to the sliced structure, the result of this approach can be seen on the right-hand side of Fig. 2.20. This, however, requires the piezo transducer stages as well as the galvanometer-driven mirrors of the lithography system to be perfectly aligned. Given this ideal alignment, structures with a smooth surface can be obtained in reasonable writing times. The inner volume of the structure can thereby be written in the fast slicing technique, whereas only the outermost shell requires slower piezo-based structuring.

The line writing technique, does not inherently guarantee every writing line to base on solid structures, such as sample surfaces or already polymerized resist for all possible PWB geometries. In a case where neither the highest nor the lowest point of a photonic wire bond lies on a chip surface or at a device facet, the conventional line writing technique fails to reliably anchor every writing line to already exposed structures or solid ground. This case is exemplarily depicted in Fig. 2.21. Precaution has to be taken in sample preparation to avoid these conditions. However, when coupling an edge emitting devices such as a laser, a semiconductor optical amplifier, or the facet of a glass fiber, to an SOI chip, tilted or ill-positioned devices are hard to avoid in manual sample preparation, given the small device dimensions.

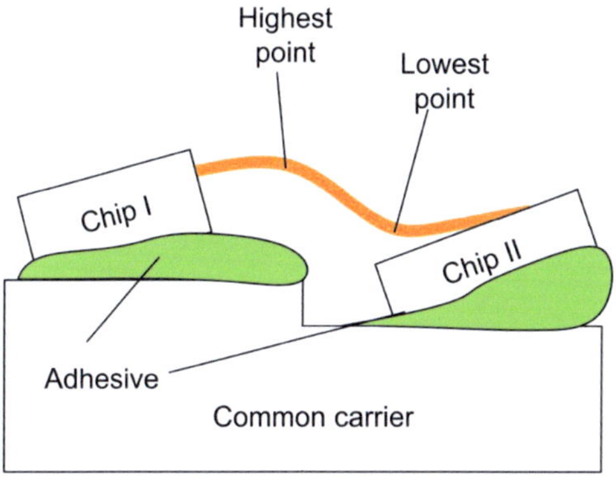

Fig. 2.21: The lowest and the highest point of a PWB connecting edge emitting devices and SOI chips do not necessary lie on the facet or surface of one of the chips. The illustration is exaggerated for clarity.

If sample tilt cannot be avoided and neither the highest nor lowest point of the PWB is securely anchored, the PWB structure has to be divided in multiple sections which are successively exposed, where each of which is attached to already written parts or chip facets or surfaces. Fig. 2.22 shows an example of how structures can be divided into multiple sections to avoid isolated writing lines. Despite this drawback the line writing strategy is the most adequate writing strategy for galvanometer-driven scanning mirror based laser lithography systems.

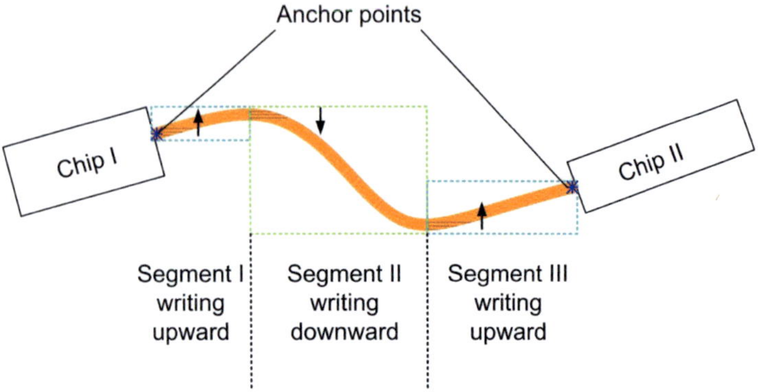

Fig. 2.22: Segmentation: The structure needs to be divided into multiple sections that are successively written. For each section the direction of slicing has to be chosen, such that every writing line bases on either a surface of the chip or an already exposed structure.

3 Connections between Silicon-on-Insulator Chips

Data rates in information processing systems continue to grow at tremendous speed and the fundamental limitations of electrical interconnects in terms of bandwidth, spatial density and power consumption are becoming increasingly obvious [57]. Due to tremendous research efforts silicon photonic platform has developed into what is currently considered the most promising option to escape this bottleneck. Silicon photonics offers high-density photonic-electronic integration with mature CMOS technology [19]. Although highly scalable devices concepts have been demonstrated in recent years, the mass production of photonic systems still stumbles. This is due to the fact that the increased integration density on-chip is at the moment not yet accompanied by an efficient and scalable off-chip interface. A breakthrough in optical packaging and interconnect technology is of crucial importance here. As a consequence the interface between an SOI nano-photonic waveguide (WG) and a photonic wire bond is of central significance for the technology platform developed in this work.

This section shows experimentally that photonic wire bonds can be fabricated from different resist materials providing low loss interconnections between silicon-on-insulators waveguides. The experimental results on the interconnections between SOI waveguides on one and on different chips are presented in this chapter.

3.1 Coupling to Nano-Photonic SOI Waveguides *via* PWB on Grating Couplers

This section reports on the first proof-of-principle experiment using a PWB coupled to SOI nanowires. The PWB presented here, has a large cross-section diameter and is coupled directly to SOI standard grating couplers [11] without the use of spot-size converters. The grating couplers are designed for off-chip connection to a standard single-mode fiber. The aim of this experiment is to proof light guiding in the *in situ* fabricated polymer waveguides as well as to give the first hint on data-transmission capabilities. The following section has been published in the *Optical Fiber Communication Conference 2011* [C21][1]. A summary of all fabrication details for the presented PWB structure is given in Appendix A.2.

– Beginning of Paper [C21] –

Photonic Waveguide Bonds – A Novel Concept for Chip-to-Chip Interconnects

N. Lindenmann, I. Kaiser, G. Balthasar, R. Bonk, D. Hillerkuss, W. Freude, J. Leuthold, C. Koos

Institute of Photonics and Quantum Electronics, Karlsruhe Institute of Technology (KIT), Engesserstr. 5, 76131 Karlsruhe, Germany

Abstract: Photonic waveguide bonds (PWB) enable three-dimensional chip-to-chip interconnects. We demonstrate for the first time a PWB link between integrated silicon-on-insulator waveguides. The viability of the concept is demonstrated by data transmission experiments.

Introduction: As data rates in information processing systems continue to grow exponentially, the fundamental limitations of electrical interconnects in terms of bandwidth, spatial density and power consumption are becoming increasingly obvious [57], [58]. Substantial research effort has been dedicated to optical interconnects, and the silicon photonic platform is currently considered the most promising option enabling high-density photonic-electronic integration with mature CMOS technology [19]. Highly scalable silicon photonic device concepts have emerged over the last years, but mass-production of optical chip-to-chip interconnect waveguides remains challenging. Inversely tapered

---

[1] The images in this sections have been reformatted for better readability, reference numbering and figure numbering have been changed for consistency.

silicon nanowires have been used for in-plane connection to single-mode fibers – either directly [13] or *via* intermediate polymer or SiO_2 waveguides [15]–[17]. Alternatively, out-of-plane fiber-chip coupling can be accomplished by grating structures that are etched into integrated waveguides [3]. Grating couplers allow optical outlets to be positioned anywhere on the chip surface and can simultaneously act as integrated polarization splitters [7]. Coupling performance of grating couplers can be improved by using bottom mirrors [3], [4], silicon overlays [5] or numerically optimized apodized grating designs [6]. Nevertheless, all these approaches still rely on active mechanical alignment and high-precision attachment of fibers and are hence unsuited for large-scale chip-to-chip interconnects in industrial mass production. While tremendous progress is being made in on-chip integration of complex photonic systems, there is currently no technology that could cope with the associated inter-chip connectivity challenges.

In this paper, we introduce and experimentally validate the concept of photonic waveguide bonds (PWB). In analogy to wire bonding of electrical integrated circuits (IC), photonic waveguide bonding paves the way towards automated fabrication of photonic chip-to-chip interconnects. The technique is based on three-dimensional (3D) in-situ structuring of negative-tone resist materials with two-photon polymerization (TPP) [25], [27], [36]. The PWB approach neither requires high-precision alignment of optical devices nor is the spatial interconnect density limited by the diameter of standard optical fibers. In a proof-of-concept experiment, we have realized for the first time PWB interconnects between silicon-on-insulator (SOI) waveguides and demonstrated their viability in data transmission experiments. Simulations indicate that the concept enables highly efficient single-mode interconnects with multi-Tbit/s transmission capacity and pitches of less than 5 µm.

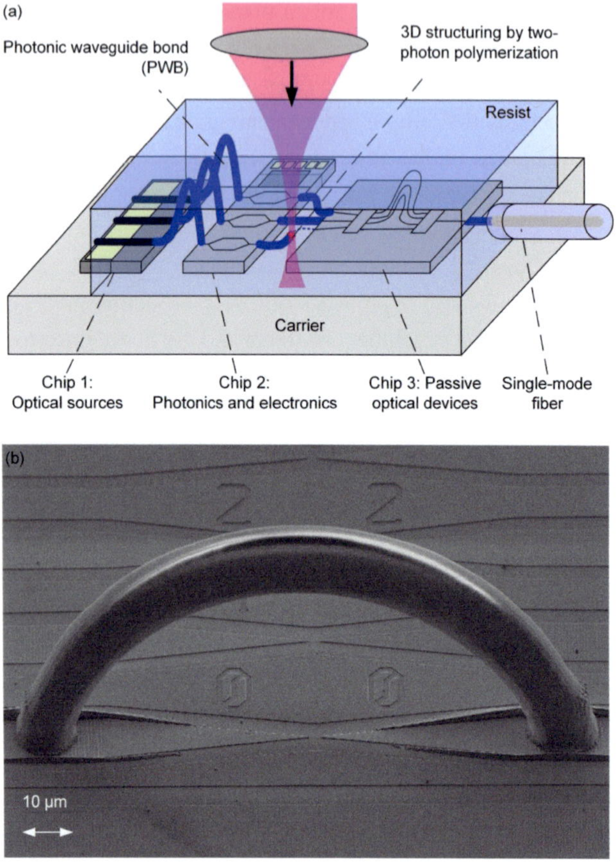

Fig. 3.1: Photonic waveguide bond (PWB) concept. (a) Photonic chips (Chip 1, Chip 2, Chip 3) and optical fiber ends are mounted on a carrier and embedded into a photosensitive negative-tone photoresist. PWB are then defined in the volume of the resist by two-photon polymerization. (b) PWB prototype fabricated in SU-8 on a chip with planar integrated SOI waveguides. The PWB arc spans a distance of 100 µm, the radius of curvature is 53 µm, and the waveguide diameter is 9 µm. Light is coupled from the planar waveguides to the PWB by grating couplers.

Photonic waveguide bond (PWB) concept: The PWB concept is illustrated in Fig. 3.1. Integrated photonic subsystems and optical fiber ends are mounted on a carrier and embedded into a photosensitive negative-tone resist material. After measuring the relative positions of the integrated waveguide facets, matched PWB geometries are computed. The waveguide structures are then defined by a direct-write 3D lithography technique which is based on two-photon polymerization in the focus of a pulsed laser beam. TPP enables the fabrication of high-index contrast chip-to-chip interconnect waveguides with almost arbitrary waveguide geometries and feature sizes as small as 100 nm [27]. The unexposed regions are removed in a development step and can subsequently be overclad with a low-index cover material. This technique enables 3D integration of complex photonic systems with high interconnect density. Since the waveguide geometry is adapted to the relative positions of the photonic subsystems, no high-precision alignment is required during fabrication and the technique is viable for fully automated industrial production.

Experimental demonstration: For a proof-of principle experiment, a PWB prototype was fabricated on an SOI chip, Fig. 3.2. b. MicroChem SU-8 2075 photoresist (ref. index n = 1.57 at 1550 nm) was deposited on pre-patterned SOI waveguides. PWB structures were defined with a commercially available 3D laser lithography system ('Photonic Professional' by Nanoscribe GmbH, [39]). Light is coupled from the planar SOI waveguides to the PWB by uniform grating couplers that were originally designed for free-space coupling to standard single-mode fibers. Since the grating patch is 12 µm wide and 15 µm long, a highly multimode PWB with a diameter of 9 µm had to be chosen. The prototype was characterized by measuring the transmission of the test structures with and without PWB, Fig. 3.1. a and b. After subtracting the wavelength-dependent losses of the SOI waveguides and grating couplers, an excess insertion loss between 7 dB and 11 dB was found for the PWB, Fig. 3.2. c. This loss is caused by mode conversion within the multimode PWB section as confirmed by simulations, Fig. 3.2. c. For single-mode PWB designs with dedicated SOI coupling structures, excess losses are expected to vanish. The PWB prototype was tested with 40 Gbit/s on-off-keying data streams, Fig. 3.3.

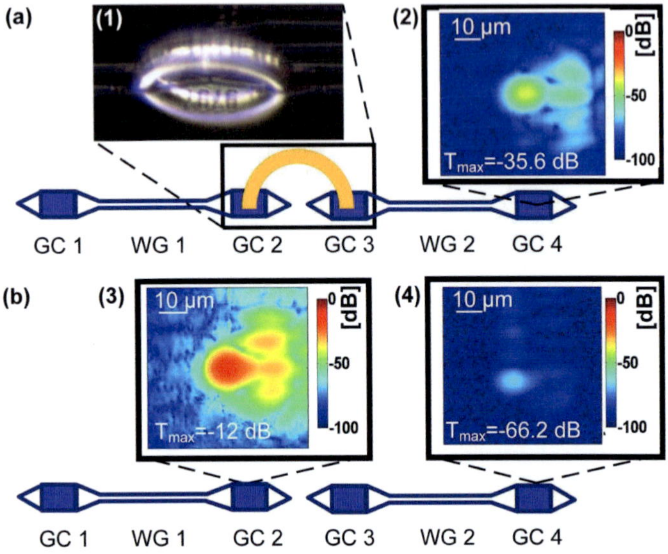

Fig. 3.2: Characterization of the fabricated PWB (a) The test structure comprises two integrated SOI waveguides (WG 1, WG 2) with grating couplers (GC 1–4). 1550 nm light is coupled from a single-mode fiber to GC 1 while GC 2 and GC 3 are connected by the PWB protoype (Inset 1). Inset 2 indicates the spatial distribution of light radiated from the waveguide end as obtained by laterally scanning a fiber across GC 4. The transmission of the test structure was measured to be -35.6 dB for optimum alignment of the output fiber. The SOI waveguides are 4 mm long, 500 nm wide and 220 nm high. (b) Spurious coupling from WG 1' to WG 2' can be neglected. To prove this, a nominally identical SOI reference structure (WG 1', WG 2') without PWB was investigated. For the first stage (GC 1' – WG 1' – GC 2'), the insertion loss for optimum alignment of the output fiber amounts to 12 dB (Inset 3), while essentially no transmission to GC 4' could be observed (Inset 4). (c) Simulated (beam propagation method, BPM) and measured transmission properties of the PWB. The transmission spectrum (blue) of the entire test structure was measured with a tunable laser. After removing the wavelength-dependent influence of the SOI waveguides (WG 1, WG 2; 2 x 2.4 dB insertion loss at 1550 nm) and the grating couplers (GC 1–4; 4 x 4.8 dB insertion loss at 1550 nm), the measured (red) and the simulated transmission (black) of the PWB agree fairly well.

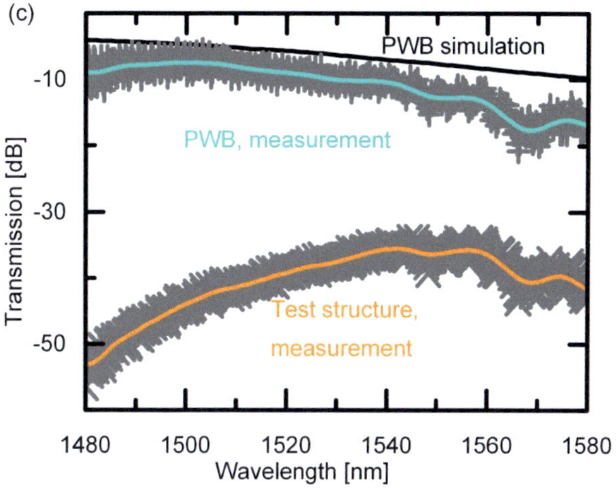

Fig. 3.2. Characterization of the fabricated PWB (c) Simulated (beam propagation method, BPM) and measured transmission properties of the PWB. The transmission spectrum (orange) of the entire test structure was measured with a tunable laser. After removing the wavelength-dependent influence of the SOI waveguides (WG 1, WG 2; 2 x 2.4 dB insertion loss at 1550 nm) and the grating couplers (GC 1–4; 4 x 4.8 dB insertion loss at 1550 nm), the measured (blue) and the simulated transmission (black) of the PWB agree fairly well.

Bit error ratios (BER) of the (2^{31}-1) pseudo-random bit sequence (PRBS) were measured as a function of receiver input power for back-to-back transmission, single-channel operation and co-propagation of two 40 Gbit/s wavelength channels, Fig. 3.3. b. The PWB does not introduce any power penalty.

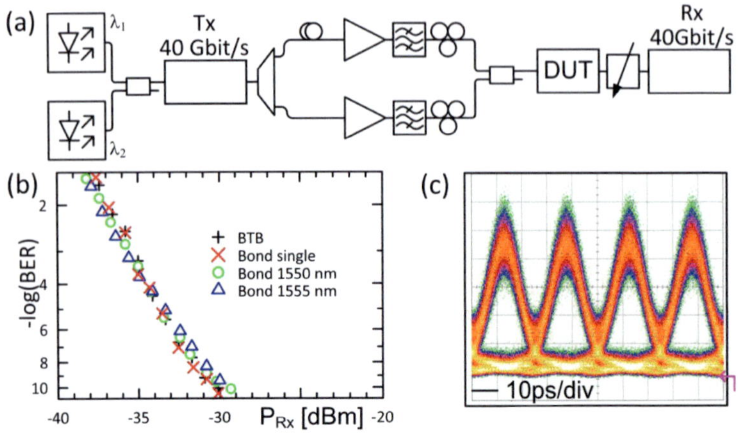

Fig. 3.3: Data transmission experiment. (a) Experimental setup: 40 Gbit/s 33% return-to-zero (RZ) on-off-keying (OOK) data streams are modulated on optical carriers at λ_1 = 1550.92 nm and λ_2 = 1555.75 nm, decorrelated by a temporal delay, amplified, and coupled into the device under test (DUT). Bit error ratios of the 2^{31}-1 PRBS are measured as a function of receiver input power after the DUT. (b) BER vs. receiver input power for back-to-back measurement, single-wavelength data transmission and co-propagation of two signals through the PWB. All curves coincide; the PWB does not exhibit any power penalty. (c) Eye diagram (λ = 1550.92 nm, BER = 10^{-9}) after the DUT for co-propagation of two wavelength channels. No nonlinear interaction is observed.

Low-loss PWB design: Excess loss caused by mode conversion can be avoided by single-mode PWB. Inversely tapered SOI structures enable efficient broadband coupling between integrated waveguides and PWB sections, Fig. 3.4. The coupling structures have been modeled with a commercially available fully vectorial time-domain solver. As one of the most crucial parameters, the tip width of the SOI waveguide taper is varied between 20 nm and 100 nm. Transmission efficiencies between 85 % and 95 % are found over the full wavelength range between 1300 nm and 1600 nm thus enabling multi-Tbit/s transmission capacity for a single PWB. Losses induced by material absorption (3 dB/cm for SU-8 at 1550 nm) can be neglected for typical PWB lengths of less than 1 mm. Interconnect pitches can be smaller than 5 µm which would correspond to tens of thousands of photonic interconnects per square millimeter.

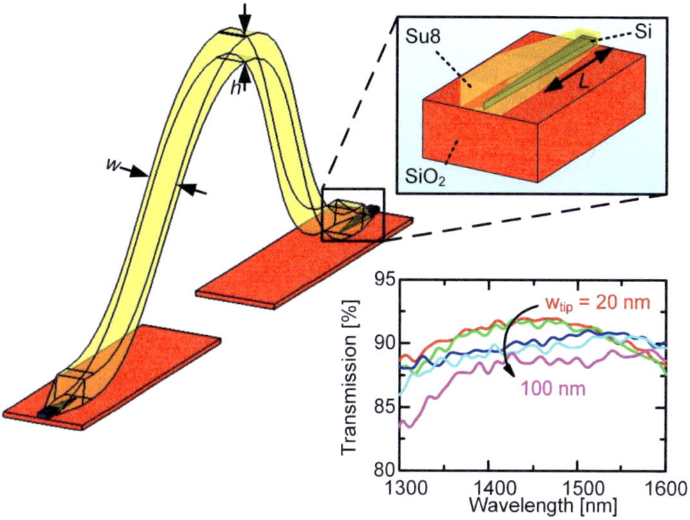

Fig. 3.4: Low-loss PWB design based on inversely tapered silicon nanowires and a SU-8 (n = 1.57 at 1550 nm) interconnect waveguide. The PWB is embedded in a low-index overcladding (Cytop, n = 1.34 at 1550 nm). The inset shows the calculated transmission efficiency of a nanowire-PWB junction for different tip widths w_{tip} of the inverse taper. Transmission efficiencies larger than 85 % (insertion losses below 0.7 dB) are found for the full wavelength range between 1300 nm and 1600 nm. At the same time, an unprecedented spatial interconnect density can be achieved: The PWB width is $w = 1.4$ µm, the height amounts to $h = 1$ µm, allowing for interconnect pitches of less than 5 µm. The tapered PWB section features a length of $L = 20$ µm; the inversely tapered SOI waveguide sections are approximately 30 µm long. Simulations were performed using CST Microwave Studio.

Summary: Photonic waveguide bonding is introduced as a novel concept for automated fabrication of intra-system 3D chip-to-chip interconnects. In a proof-of-principle experiment, we have demonstrated for the first time a PWB link between nanophotonic SOI waveguides. The viability of the concept is shown in a data transmission experiment. Simulations indicate the feasibility of multi-Tbit/s interconnects with unprecedented spatial density.

– End of paper [C21] –

This experiment shows that photonic wire bonds can be used as interconnect waveguides in data communication. The high measured insertion losses of -7 to -11 dB are expected to significantly improve by more advanced coupling structures as well as by the use of PWB waveguides with single mode cross-sections.

3.2 Chip-to-Chip Interconnections

Due to the rather big emission spot at the grating couplers, direct coupling between photonic wire bonds and grating couplers is not considered as a favorable coupling scenario for PWB to chip interfaces. Moreover, the use of grating couplers leads to relatively narrow transmission bandwidths that are undesired for broadband optical communication. Having shown the proof-of-principle experiment for on-chip multimode PWB interconnects in the previous section, the following section deals with the simulation of advanced taper structures as well as with the experimental results on the fabrication and characterization of inter-chip PWB connections.

3.2.1 The Three-Dimensional Double Taper Approach

Inversely tapered silicon nanowires, in direct contrast to grating couplers, can be used as broadband mode converters. They have been used for in-plane connection to single-mode fibers – either directly [13] or *via* intermediate polymer or SiO_2 waveguides [15]–[17]. A similar approach is followed here. Since the respective mode field diameters of the SOI and PWB ground modes are very different, as can be seen in Fig. 3.5, tapering has to be applied for a low-loss, broadband and polarization insensitive transition between PWB and SOI waveguide. Laterally down-tapering the SOI WG towards the PWB ("inverse taper") [13] is exploited. The SOI waveguide has a width w_{Si} in the un-tapered region. Over a length l_{Si}, the SOI waveguide is tapered down to a final width w_{tip}. The waveguide height h_{Si} is constant throughout the whole waveguide.

The tapered SOI WG is embedded into a polymer waveguide that is down-tapered towards the SOI WG as is depicted in Fig. 3.6. The polymer waveguide in the untapped case has a width $w_{PWB,taper,2}$ and a height $h_{PWB,taper,2}$. Over a length $l_{PWB,taper}$ the polymer waveguide is tapered down to a width $w_{PWB,taper,1}$ and a height $h_{PWB,taper,1}$. This structure, comprising SOI and PWB down-tapers is referred to as a 3D double-taper in the following. The 3D double-taper is covered by a lower index cladding material that simultaneously acts as a cladding material for the photonic wire bond.

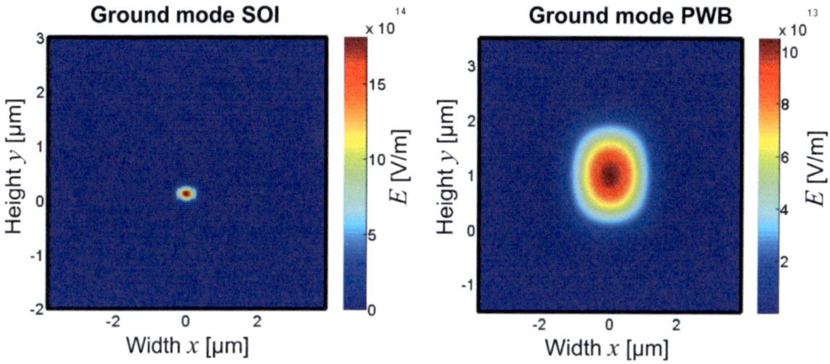

Fig. 3.5: Ground modes of an SOI waveguide consisting of a 500 nm wide and 220 nm high silicon strip (refractive index n_{Si} = 3.48 at 1550 nm) deposited on a silicon dioxide buffer layer (n_{SiO2} = 1.44 at 1550 nm) and a rectangular PWB waveguide (*IP-Dip*™ [39], $n_{IP\text{-}Dip}$= 1.53 at 1550 nm) with a cross-section of 2.8 µm × 2 µm embedded into a lower index cladding material e. g., *Cytop*™[59], n_{Cy} = 1.34 at 1550 nm. The total energy in the mode is normalized to 1J [60].

In the presented model, the SOI waveguide consists of a 500 nm wide and 220 nm high silicon strip (refractive index n_{Si} = 3.48 at 1550 nm) deposited on a silicon dioxide buffer layer (n_{SiO2} = 1.44 at 1550 nm). The PWB waveguide core is made of a polymer material (SU-8 [38], n_{SU8} = 1.57 at 1550 nm or *IP-Dip*™[39], $n_{IP\text{-}Dip}$ =1.53 at 1550 nm) and features a rectangular cross-section at the interface towards the PWB waveguide section. The PWB waveguide core is immersed into a low-index cladding (e. g., *Cytop*™ [59], n_{Cy} = 1.34 at 1550 nm), which is not depicted in Fig. 3.6 for the sake of clarity. Absorption losses in the polymer waveguide of the PWB can usually be neglected. In order to obtain design guidelines for the 3D double-taper numerous simulations on the transition between the integrated SOI waveguide and the PWB section are performed, the results of which are summarized in Appendix A.2

As a result, we identify the silicon taper width as the most crucial influence factor on the performance of the double-taper structure. While a change of the PWB taper width and height of several hundred nanometers only results in additional losses of less than 0.1 dB compared to the optimum, the coupling efficiency of the 3D double taper structure drops by 1.2 dB when increasing the taper tip width from $w_{tip} = 80$ nm to $w_{tip} = 200$ nm. It is therefore crucial to use smallest possible taper tip width w_{tip} in order to obtain reasonable optical transmission. Due to practical reasons, however, taper tip width smaller than $w_{tip} = 80$ nm are not available from foundry services at the present stage.

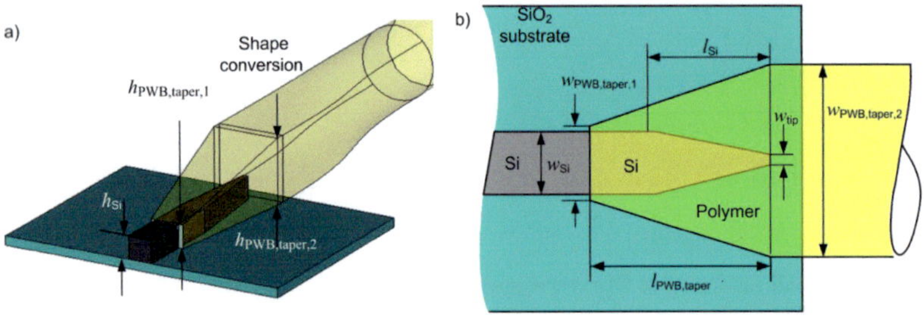

Fig. 3.6: 3D Double-taper structure for modal field match between PWB and SOI waveguide. (a) The SOI WG ends in a laterally down-tapered section ("inverse taper"), whereas the PWB features a section that is laterally and vertically down-tapered to adiabatically transform the SOI waveguide mode to that of the PWB. The PWB waveguide section gradually changes its cross-section from rectangular to circular. (b) The SOI WG tip has a width of $w_{tip} = 80$ nm. The SOI WG taper connects to a straight SOI WG with a cross-section of $w_{Si} = 500$ nm by $h_{Si} = 220$ nm. The lower-index cladding material is not displayed here for the sake of clarity.

3.2.2 Chip-to-Chip Prototype

In this section, a PWB that bridges the gap between two different silicon photonic chips is presented. Using first, however, un-optimized inversely-tapered transitions[2] between silicon photonic waveguides and photonic wire bond section, light transmission over a spectral width of 100 nm is demonstrated. This work has been published on the *IEEE International Conference on Group IV*

[2] SOI chips with optimized taper structures have not been present at the early stage of the project.

Photonics 2011 [C20][3]. The fabrication details for the devices demonstrated in this section are summarized in Appendix A.3

– Beginning of Paper [C20] –

Photonic Wire Bonding for Single-Mode Chip-to-Chip Interconnects

N. Lindenmann[1], G. Balthasar[1], R. Palmer[1], S. Schuele[2], J. Leuthold[1,2], W. Freude[1,2], C. Koos[1,2]

[1] *Institute of Photonics and Quantum Electronics (IPQ), Karlsruhe Institute of Technology (KIT), Engesserstr. 5, 76131 Karlsruhe Germany*

[2] *Institute of Microstructure Technology (IMT), Karlsruhe Institute of Technology (KIT), Hermann-von-Helmholtz-Platz 1,76344 Eggenstein-Leopoldshafen, Germany*

Abstract: Photonic wire bonds (PWB) enable single-mode chip-to-chip interconnects that are suitable for mass production. We demonstrate for the first time a single-mode PWB link between two different nanophotonic silicon-on-insulator chips.

Introduction: Due to ever-increasing data rates in information processing systems, the fundamental limitations of electrical interconnects in terms of bandwidth, spatial density and power consumption become increasingly obvious [57], [58]. Driven by substantial world-wide research investments, tremendous progress has been made in the field of photonic integration over the last years. The silicon photonic platform is currently considered the most promising option [19] enabling high-density photonic-electronic integration by employing mature CMOS technology. However, there is currently no technology that could cope with the associated photonic inter-chip connectivity challenges

[3] The images in this sections have been reformatted for better readability, reference numbering and figure numbering have been changed for consistency.

which is the main obstacle towards industrial adoption of large-scale photonic-electronic integration.

Current research mainly concentrates on refined fiber-chip coupling techniques. Inversely tapered silicon nanowires were used for in-plane connection to single-mode fibers – either directly [13], or *via* intermediate polymer or SiO_2 waveguides [15]–[17]. Alternatively, out-of-plane fiber-chip coupling can be accomplished by grating structures that are etched into integrated waveguides [3]. Grating couplers allow optical outlets to be positioned anywhere on the chip surface and can simultaneously act as integrated polarization splitters [7]. Coupling performance of grating couplers can be improved by using bottom mirrors [3], [4], silicon overlays [5] or numerically optimized apodized grating designs [6]. However, all these approaches still rely on active mechanical alignment and high-precision attachment of fibers and are hence unsuited for large-scale chip-to-chip interconnects in industrial mass production.

In this paper we target a novel technology for single-mode chip-to-chip interconnects. The technology is based on in-situ fabrication of three-dimensional freeform waveguides between prepositioned chips. It does neither require highly precise active alignment of optical chips, nor is the interconnect density limited by the diameter of standard single-mode fibers. In analogy to wire bonding of electrical integrated circuits, we refer to our method as photonic wire bonding. We have recently demonstrated the fabrication of a multi-mode photonic wire bond waveguide in conjunction with grating couplers both of which were situated on the same silicon-on-insulator chip [C21]. In this paper we demonstrate for the first time a single-mode photonic wire bond that bridges the gap between two different silicon photonic chips. Using inversely-tapered transitions between silicon photonic waveguides and photonic wire bond sections, we demonstrate light transmission over a spectral width of 100 nm. We consider this work to be a major breakthrough towards highly automated production of photonic chip-to-chip interconnects without the need for high-precision mechanical alignment.

Fabrication Concept: In situ fabrication of photonic wire bonds (PWB) makes use of direct-write 3D lithography based on two-photon polymerization in the focus of a pulsed laser, Fig. 3.7. a.

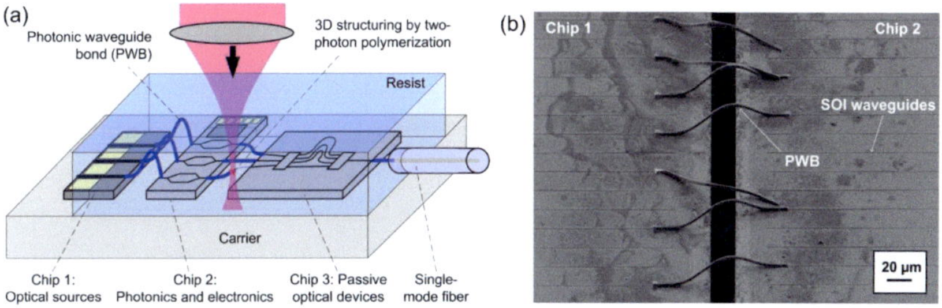

Fig. 3.7: Photonic wire bonding concept. (a) Photonic chips (Chip 1, Chip 2, Chip 3) and optical fiber ends are fixed to a common carrier and embedded into a negative-tone photoresist. Photonic wire bonds (PWB) are then defined in the volume of the resist by two-photon polymerization. (b) PWB prototypes fabricated between two planar integrated optical chips. The PWB arcs span a distance of approximately 80 µm; the gap between the chips is ~ 20 µm wide. Some of the PWB have detached from the chip during development of the photoresist.

The entire fabrication process comprises several steps: First silicon-on-insulator (SOI) chips with processed waveguide structures are mounted on a common carrier. These chips already contain coupling structures such as inverse tapers, grating couplers as well as reference marks. The assembly is then covered with the negative-tone photoresist, and the spatial positions of the coupling structures are captured by a highly precise 3D metrology system. The relative positions of the SOI waveguide facets are then used to calculate the three-dimensional geometry of the PWB waveguide. The PWB structures are then defined by direct-write 3D lithography. In the next step the unexposed resist material is removed in a development step and the sample is finally cladded with a lower index cover material. This cover material acts as a mechanical support of the free-standing PWB waveguide cores and ensures single-mode operation with lateral waveguide dimensions larger than 1 µm. Since the waveguide geometry is adapted to the position of the integrated waveguides, our approach does not require high-precision alignment of optical chips. Photonic wire bonding therefore enables high-throughput mass-production of PWB with fully automated systems.

Experimental Demonstration: We have fabricated a single-mode PWB that connects two nanophotonic SOI waveguides on different chips, Fig. 3.8. For this experiment a PWB inter - chip interconnect waveguide was fabricated on two nanophotonic silicon on insulator chips. The polymer waveguide core has a nearly rectangular cross section of approximately 2 µm by 1.6 µm. It consists of Su8, having a refractive index of 1.56 at a wavelength of 1550 nm. Index matching liquid (n = 1.33 at 1550 nm) is used as a cladding material. Optical coupling between the SOI waveguides and the PWB sections is achieved by inversely tapered SOI waveguides which are combined with three-dimensional taper structures in the polymer. The total length of the taper structures is approximately 20 µm; the tip widths of the inverse SOI taper amounts to 100 nm.

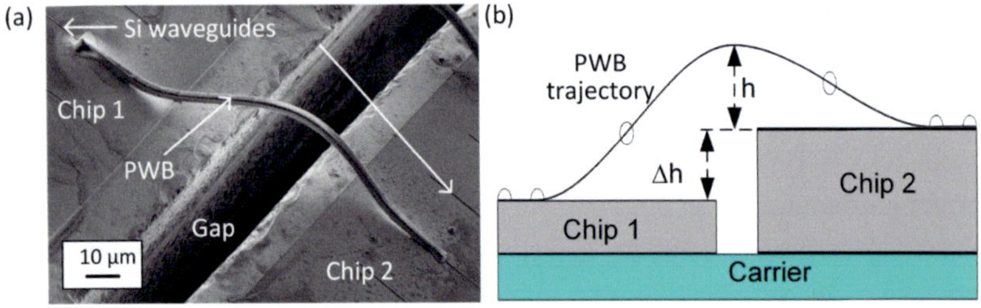

Fig. 3.8: (a) Fabricated photonic wire bond (PWB). The bond has a diameter of approximately 2 µm. The polymer waveguide core consists of Su8, having a refractive index of 1.56 at a wavelength of 1550 nm. Index matching liquid (n = 1.33 at 1550 nm) is used as a cladding material. (b) Lateral view of the PWB prototype. The PWB compensates for a height difference of Δh = 12 µm between the surface of chip 1 and chip 2. The height of the apex point amounts to $h+\Delta h$ = 30 µm. The PWB waveguide has a nearly rectangular cross section of approximately 2 µm by 1.6 µm. Imperfections of the waveguide structure are due to imperfect development process.

The PWB compensates for a displacement of the two SOI chips both in the lateral and in the vertical direction. The middle section of the PWB waveguide follows the trajectory of a fourth-order polynomial in 2 dimensions with the coupling regions lying in the extrema to ensure a smooth transition between the direction of the SOI waveguides and the PWB. The height of the PWB apex (h+Δh) amounts 30 µm, Fig. 3.8. b. This is a first proof-of-principle experiment with large potential of further improving waveguide fabrication and transmission efficiency.

Characterization Results: To evaluate the excess loss of our PWB prototype, we measured the transmission characteristic through a PWB-SOI chip-to-chip interconnect, Fig. 3.9. a. The transmission was normalized to a reference SOI waveguide comprising two grating couplers on a single chip. Within the considered wavelength range, the smallest observed excess losses amount to approximately 5 dB. The pronounced slope in transmission is believed to be caused by a wavelength-dependent propagation loss of the access waveguides, which are 220 nm high and only 300 nm wide.

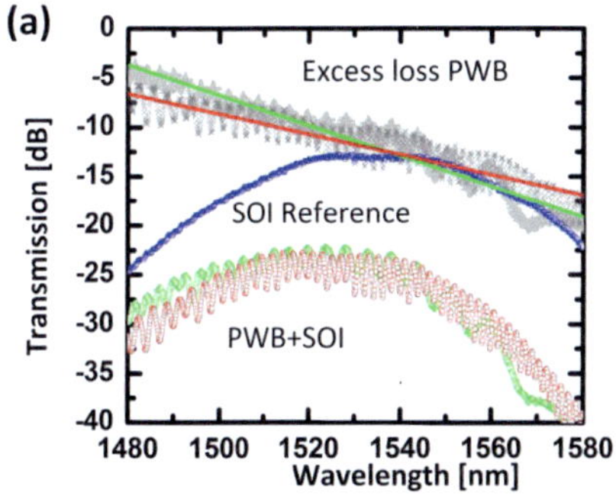

Fig. 3.9: (a) Excess loss of two different PWB (red and green) as a function of wavelength. “PWB + SOI” denotes the transmission through a PWB-SOI chip-to-chip interconnect, “SOI Reference” refers to a reference SOI waveguide comprising two grating couplers on a single chip, as depicted in (b). The excess loss of the PWB is obtained by normalizing the PWB+SOI-transmission to the SOI Reference (blue).

Strong Fabry Perot fringes are caused by back reflections within the high index-contrast silicon-on-insulator waveguides. Excess losses can be considerably reduced by optimized fabrication processes.

Summary: Photonic wire bonding is introduced as a novel concept for automated fabrication of 3D freeform single-mode optical chip-to-chip interconnects. We demonstrate for the first time a single-mode PWB link between two different nanophotonic SOI chips. Inverse tapers enable light transmission over a spectral width of more than 100 nm. We expect photonic wire bonding to become a key enabler for the assembly of highly scalable photonic chips.

– End of Paper [C20] –

3.2.3 Loss Reduction by Double Taper Approach

In this section PWB connecting to optimized tapers, which significantly reduces the insertion loss down to 1.6 dB in the optical C-Band (1530 nm – 1565 nm), are presented. This enables for the first time broadband single-mode transmission over the wavelength range from 1270 nm to 1580 nm.

To further increase the device performance, photonic wire bonds with improved coupling regions, based on optimized 3D double tapers have been investigated. Inversely tapered SOI structures enable efficient broadband coupling between integrated waveguides and PWB sections, which can be seen from simulations. The coupling structures have been modeled with a commercially available fully vectorial time-domain solver [61]. As one of the most crucial parameters, the tip width of the SOI waveguide taper is varied between 20 nm and 100 nm. Transmission efficiencies between 85 % and 95 % are found over the full wavelength range between 1300 nm and 1600 nm as can be seen in Fig. 3.10. This broad transmission range results in multi-Tbit/s transmission capacity for a single PWB. Losses induced by material absorption (3 dB/cm for SU-8 [38] at 1550 nm) can be neglected for typical PWB lengths of less than 1 mm. More simulation results on the influence of taper dimensions can be found in Appendix A.4. Interconnect pitches can be smaller than 5 μm which would correspond to tens of thousands of photonic interconnects per square millimeter.

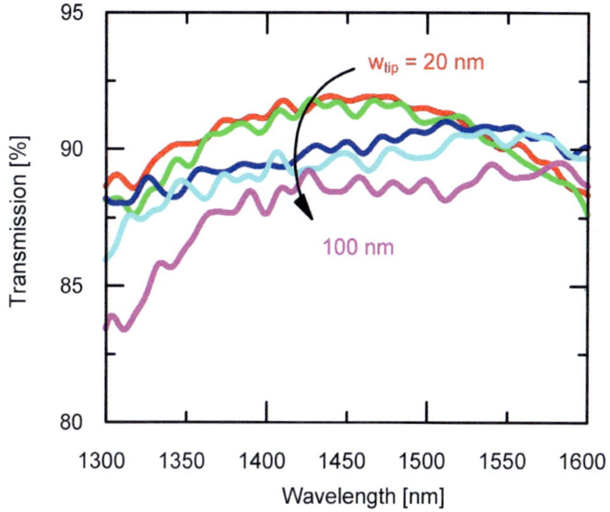

Fig. 3.10: Calculated transmission efficiency of a nanowire-PWB junction for different tip widths w_{tip} of the SOI down-taper. Transmission efficiencies larger than 85 % (insertion losses below 0.7 dB) are found for the full wavelength range between 1300 nm and 1600 nm not taking into account material dispersion. At the same time, an unprecedented spatial interconnect density can be achieved: The PWB width is $w_{PWB,taper,2} = 1.4$ µm, the height amounts to $h_{PWB,taper,2} = 1$ µm in this simulation. The taper tip width and height of the PWB taper are $w_{PWB,taper,1} = 0.76$ µm by $h_{PWB,taper,1} = 0.45$ µm. The tapered PWB section features a length of $l_{PWB,taper} = 20$ µm; the inversely tapered SOI waveguide sections are approximately 30 µm long. Simulations were performed using CST Microwave Studio.

The device preparation and lithography parameters are summarized in Appendix A.4. In contrast to the previous experiment the port detection algorithm described in Section 2.1 has been exploited here resulting in enhanced placement accuracy. The trajectory of the PWB waveguide follows a fifth-order polynomial in two dimensions which is designed to provide a smooth transition between the SOI waveguides and the PWB.

The PWB fabricated for this experiment connect two tapered SOI waveguides, see Fig. 3.11. The SOI waveguide in the un-tapered region has a width of $w_{Si} = 500$ nm and a height $h_{Si} = 220$ nm. The tip width of the inverse SOI taper amounts to $w_{tip} = 80$ nm. The length of the PWB taper structure is approximately $l_{PWB,taper}$ = 100 µm. The PWB waveguide core has a nearly rectangular cross-section of approximately $w_{PWB,taper,2}$ =2 µm by $h_{PWB,taper,2}$ = 1.6 µm. It is fabricated from SU-8 [38], which has a refractive index of 1.57 at a wavelength of 1550 nm. Index matching liquid (n = 1.3 at 1550 nm) is used as a cladding material. The PWB taper tip width and height are designed to be $w_{PWB,taper,1}$ = 0.76 µm by $h_{PWB,taper,1}$ = 0.45 µm.

The photonic wire bond is fabricated in the trajectory-writing approach presented in Section 2.4.1. The piezo-based lithography setup described in Section 1.3.3 is used. A SEM image of the fabricated structures can be seen in Fig. 3.11.

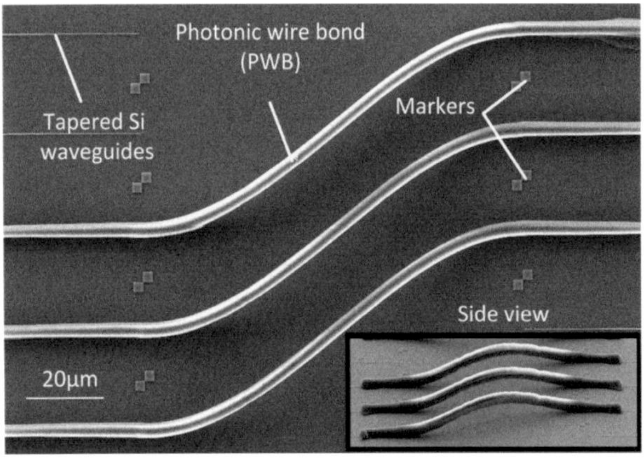

Fig. 3.11: PWB connecting inversely tapered silicon-on-insulator waveguides. The PWB core has a nearly rectangular cross-section of approximately 2 µm by 1.6 µm and connects two SOI nanowires with a 500 nm × 200 nm cross-section. The PWB waveguide is made of SU-8 [38] with a refractive index of 1.57 at a wavelength of 1550 nm. The middle section of the PWB waveguide follows the trajectory of a fourth-order polynomial in two dimensions which is designed to provide a smooth transition between the SOI waveguides and the PWB.

Experimentally, the photonic wire bond prototypes show low-loss and broad optical bandwidth, as can be seen in Fig. 3.13. The measurement is performed using the setup depicted in Fig. 3.12. We use a tunable laser source together with an optical spectrum analyzer to cover the infrared telecommunication wavelength range between 1270 nm and 1580 nm. Standard single-mode fibers in combination with conventional grating structures are used to couple light to and from the chip. To obtain the insertion loss of the PWB, the spectrum of the PWB together with SOI feed waveguides is first measured. The result is then normalized to the transmission spectrum of a same-length silicon reference waveguide on the same chip to isolate the contribution of the wire bond. Due to the behavior of the grating structures, the absolute transmission in the wavelength range below 1400 nm is small which leads to a more pronounced noise influence in the measurement between 1270 nm and 1400 nm. The PWB's average insertion loss in the C band (1530 nm–1565 nm) amounts to 1.6 dB with a standard deviation of 0.13 dB. For the whole wavelength range between 1270 nm and 1580 nm insertion losses of 2.5 dB with a standard deviation of 1.1 dB are found, see Fig. 3.13.

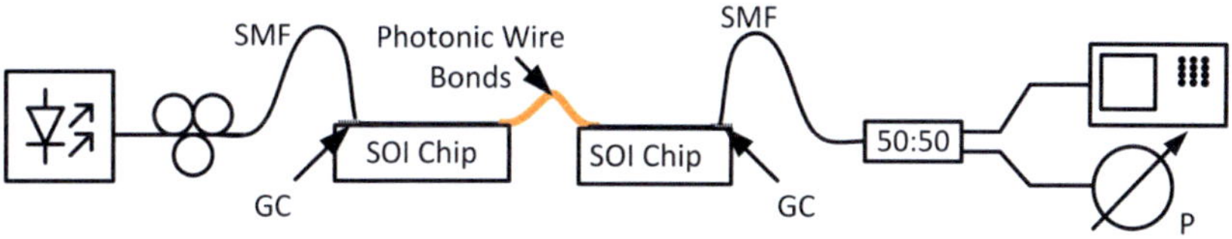

Fig. 3.12: Measurement setup for the characterization of chip-to-chip PWB. Light from a tunable laser source is coupled to the SOI chip by means of grating couplers (GC). A polarization controller is used to choose the input polarization. Light is transmitted through the photonic wire bond to the second SOI chip. From a second grating coupler it is launched back into the optical fiber. It is split up in a 50:50 coupler and detected by an optical spectrum analyzer as well as a power meter.

Due to the broadband and low-loss transmission characteristics, the PWB demonstrated in the above-mentioned experiment are highly suited for multi Tbit/s data transmission which has been proven experimentally and is discussed in the next section.

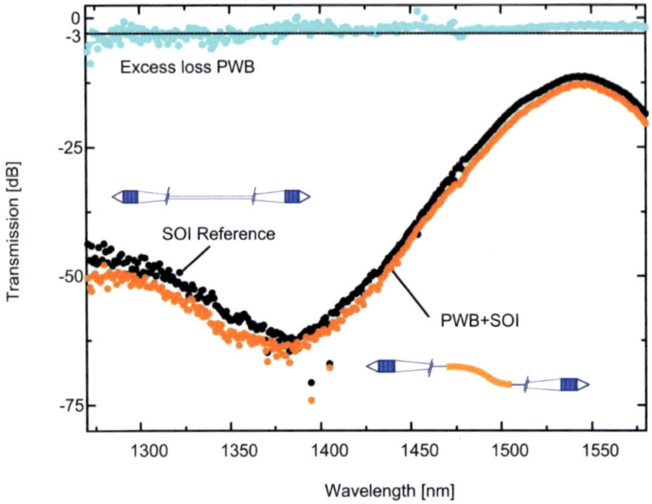

Fig. 3.13: Transmission measurement of a PWB (blue) connecting two tapered nano-photonic silicon-on-insulator waveguides. The measurement is normalized to an SOI reference waveguide (black) on the same chip, the measured transmission through the whole assembly consisting of SOI waveguides, grating couplers and the PWB is depicted in orange. The 0 dB and 3 dB loss lines are depicted in black. The average loss in the C-band (1530-1565 nm) amounts to 1.6 dB.

3.3 Multi-TBit/s Data Transmission Through Photonic Wire Bonds

To prove that PWB not only offer low-loss light transmission but also provide excellent data transmission properties, a high data-rate transmission experiment was performed and the signal quality with and without photonic wire bond was measured, finding no measurable signal degradation.

The following section has been published in Proceedings of the *Optical Fiber Communication 2012* [C16][4]. The device under test used here has been introduced in Section 3.2.3. All fabrication details for the devices used in this section are summarized in Appendix A.4.

---

[4] reference numbering and figure numbering have been changed for consistency.

– Beginning of Paper [C16] –

Low-Loss Photonic Wire Bond Interconnects Enabling 5 TBit/s Data Transmission

N. Lindenmann[1], G. Balthasar[1], M. Jordan[1], D. Hillerkuss[1,]
R. Schmogrow[1], W.Freude[1,2], J. Leuthold[1,2], C. Koos[1,2]

[1] *Institute of Photonics and Quantum Electronics (IPQ), Karlsruhe Institute of Technology (KIT), 76131 Karlsruhe, Germany*
[2] *Institute of Microstructure Technology (IMT), Karlsruhe Institute of Technology (KIT), 76344 Eggenstein-Leopoldshafen, Germany*

Abstract: Photonic wire bonding enables flexible single-mode chip-to-chip interconnects with average losses of only 2.5 dB over a spectral range from 1270 nm to 1580 nm. Flawless transmission is demonstrated for a 5.25 Tbit/s data stream.

Introduction: To keep pace with the persistent growth of data rates in information processing systems, the fundamental limits of electrical interconnects in terms of bandwidth, spatial density and power consumption need to be overcome [57], [58]. Optical interconnects can deal with these challenges, with the silicon photonics platform being the prime candidate for realizing integrated Tbit/s optical transceiver systems with mature CMOS technology [19], [62]. On-chip integration has been demonstrated for a multitude of silicon photonic devices over the last years, but connecting nanophotonic circuits to the outside world remains challenging. Current fiber-chip coupling techniques rely either on in-plane coupling using tapered waveguides as spot-size converters [13], [15], or on out-of plane coupling by grating structures [3]–[5], [7]. However both technologies require very precise passive or active alignment of fibers with respect to the components. In addition, spatial interconnect density is usually limited by the diameter of optical fibers. The existing technologies are hence unsuited for fully automated fabrication of single-mode chip-to-chip interconnects and cannot cope with the connectivity challenges that are imposed by large-scale photonic-electronic integration.

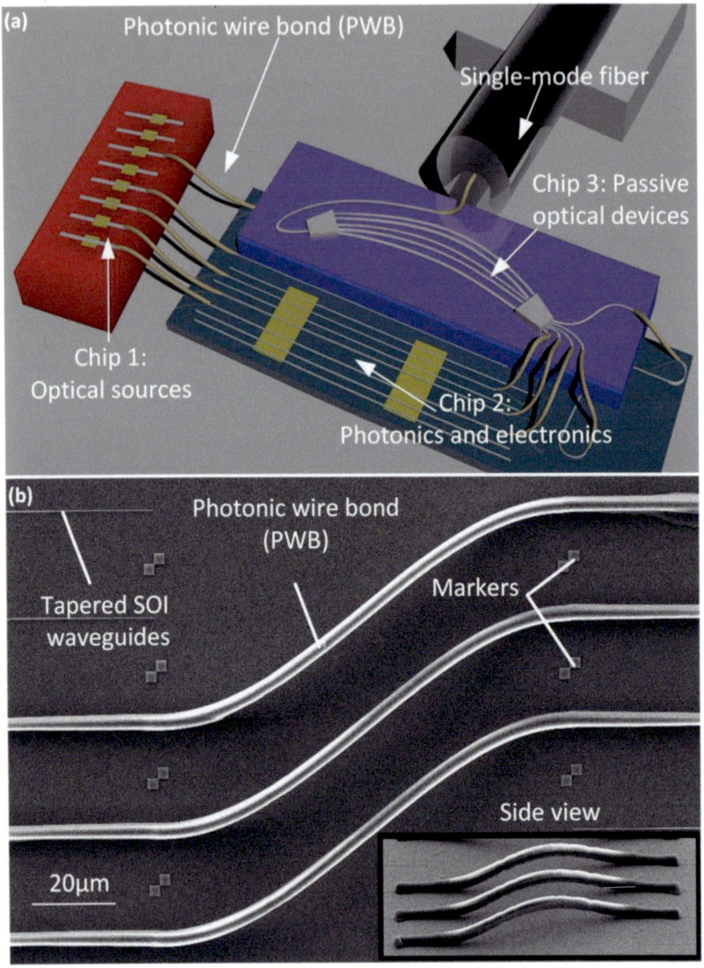

Fig. 3.14: Photonic wire bond concept. (a) Vision of PWB chip-to-chip interconnects; different photonic devices on several substrates are connected by PWB. (b) PWB connecting inversely tapered silicon-on-insulator waveguides. The PWB core has a nearly rectangular cross section of approximately 2 µm by 1.6 µm and connects two SOI nanowires with a 500 nm × 200 nm cross section. The PWB waveguide consists of SU-8 with a refractive index of 1.56 at a wavelength of 1550 nm. The middle section of the PWB waveguide follows the trajectory of a fourth-order polynomial in two dimensions which is designed to provide a smooth transition between the SOI waveguides and the PWB.

We have recently introduced photonic wire bonds as a novel approach for chip-to-chip interconnects [C21], [C20]. The technique is based on in-situ structuring of interconnect waveguides in the volume of a photosensitive resist material by means of two-photon polymerization [30]. We demonstrated the use of PWB to interconnect different nanowires *via* surface gratings and inverse tapers. In this paper we present PWB connecting to optimized tapers, which significantly reduces the insertion loss. This enables for the first time broadband single-mode transmission over the entire wavelength range from 1270 nm to1580 nm. The viability of the concept for transmitting advanced modulation formats without distortions through reflection, PMD or nonlinear impairments is demonstrated by means of a data transmission experiment using a wavelength division multiplexed (WDM) signal with more than 100 channels, each of which carrying a 16QAM signal at a symbol rate of 12.5 GBd.

Experimentally, the photonic wire bond prototypes show low-loss and broad optical bandwidth. The measurement was performed using tunable laser sources together with an optical spectrum analyzer to cover the infrared telecommunication wavelength range between 1270 nm and 1580 nm. Standard single-mode fibers in combination with conventional grating structures are used to couple light to and from the chip, Fig. 3.15. b. To obtain the insertion loss of the wire bond, we first measured the spectrum of the PWB together with SOI feed waveguides. The result was then normalized to the transmission spectrum of a same-length silicon reference waveguide on the same chip to isolate the contribution of the wire bond. Due to the behavior of the grating structures the absolute transmission in the wavelength range below 1400 nm is small which leads to a more pronounced noise influence in the measurement between 1270 nm and 1400 nm. The PWB's average insertion loss in the C band (1530 nm–1565 nm) amounts to 1.6 dB with a standard deviation of 0.13 dB. For the whole wavelength range between 1270 nm and 1580 nm insertion losses of 2.5 dB with a standard deviation of 1.1 dB are found, Fig. 3.15. a.

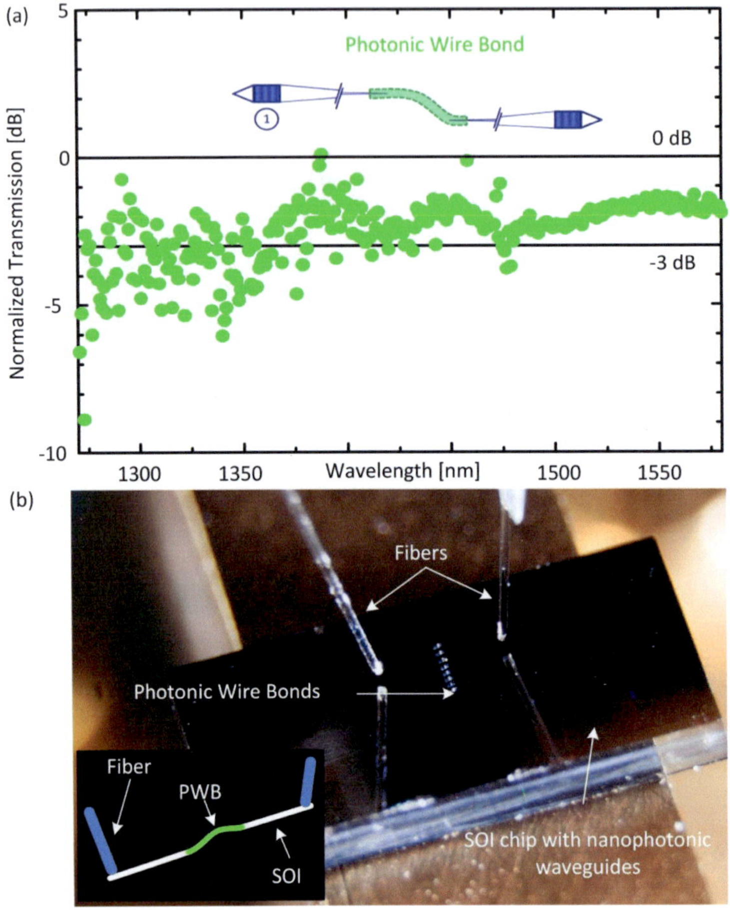

Fig. 3.15: Characterization measurement (a) Transmission measurement of a PWB (green) connecting two tapered nanophotonic silicon on insulator waveguides. The measurement is normalized to an SOI reference waveguide on the same chip. The 0 dB and 3 dB loss lines are depicted in black. The average loss in the C-band (1530-1565 nm) amounts to 1.6 dB. (b) Photograph of the SOI chip carrying the nanophotonic waveguides and the PWB. The optical coupling to the silicon waveguides is performed with standard single mode fibers and grating couplers under an angle of 80° to the sample surface.

Data transmission measurement: High-speed data transmission through a PWB link has been investigated to show the viability of the concept for Tbit/s optical interconnects. The experimental setup is depicted in Fig. 3.16. For the data transmission measurement a pulse train from a mode locked laser (MLL) is spectrally broadened in a highly nonlinear fiber (HNLF). Afterward it is shaped and sliced in a wavelength selective switch (WSS) to form a flat optical comb. This results in 105 optical carriers that cover the spectral range between 1543.42 nm and 1555.74 nm with a spacing of 12.5 GHz. An optical disinterleaver (DIL) separates odd and even carriers. Each carrier is subsequently modulated with 16QAM signals at a data rate of 12.5 GBd. In contrast to previous experiments with a similar setup [63], the signals are pulse shaped to generate rectangular spectra [64] to avoid spectral overlap of neighboring channels. The data stream with an aggregate data rate of 5.25 Tbit/s is then transmitted through the DUT. On the receiver side, the signal is detected and analyzed using an optical modulation analyzer (OMA – Agilent N4391A). We tested optical transmission in the back-to-back (BTB) case and compared it to transmission through the PWB assembly and through the SOI reference waveguide. The constellation diagram of the PWB, reference device and BTB are typical and are measured at the carrier closest to 1550 nm, see Fig. 3.16.

The EVM for almost all carriers is below the threshold for the second-generation FEC (EVM = 11.5 % for 16QAM $\approx$ BER 2.3×10^{-3}) [15]. In the BTB-case, the average EVM is 9.06 %. The transmission through the PWB as well as through the silicon reference device results in a negligible EVM penalty of 0.3 % and 0.4 %, respectively.

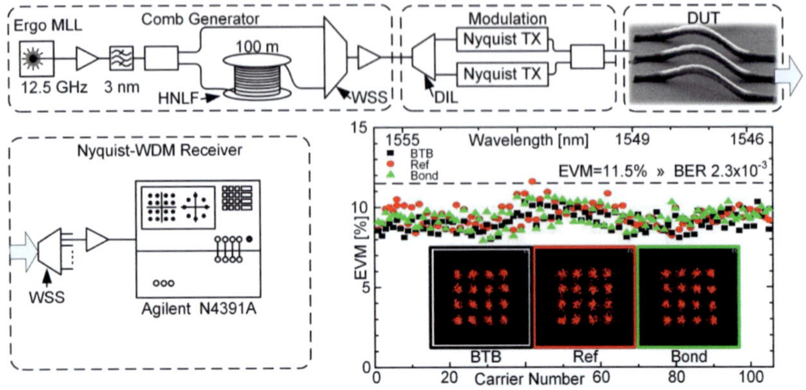

Fig. 3.16: Data transmission experiment: A number of 105 optical carriers with a spacing of 12.5 GHz between 1543.42 and 1555.74 nm are generated in an optical comb generator based on spectral broadening of a pulse train from a mode locked laser (MLL) in a highly nonlinear fiber (HNLF). After spectral shaping and slicing in a wavelength selective switch (WSS), odd and even carriers are separated in an optical disinterleaver (DIL). The carriers are subsequently encoded with Nyquist pulse shaped 16QAM signals at 12.5 GBd. After recombining odd and even carriers, the signal is transmitted through a silicon reference device or through two silicon waveguides connected by a PWB and received in a coherent-WDM Receiver with optical pre-amplifier. To gain a reference for the overall system performance, back-to-back (BTB) measurement results are provided. The shown constellation diagram for transmission through the PWB, the silicon reference device and in the BTB case are typical and are measured at the carrier closest to 1550 nm. The EVM for almost all carriers is below the threshold for second generation FEC (BER 2.3×10^{-3}).

Photonic wire bonding has been demonstrated to enable broadband low-loss coupling between single-mode silicon-on-insulator waveguides. We demonstrate the first time multi-Tbit/s data transmission through a photonic wire bond, reaching data rates of 5.25 TBit/s without measurable signal degradation. We expect photonic wire bonding to become a key enabler for the assembly of highly scalable multi-chip photonic systems.

– End of Paper [C16] –

3.4 Photonic Wire Bonds fabricated from Ormocer Material

The photonic wire bonding concept can be extended to different resist materials including dedicated waveguide materials. This is demonstrated by using Ormocer®-I [65], [66] for the fabrication of several nominally identical PWB on an SOI chip. This material is a so-called organically modified ceramic material developed by the Fraunhofer IZM in Würzburg, Germany. It provides a high transparency at the wavelength of 1550nm, as can be seen in Fig. 3.17.

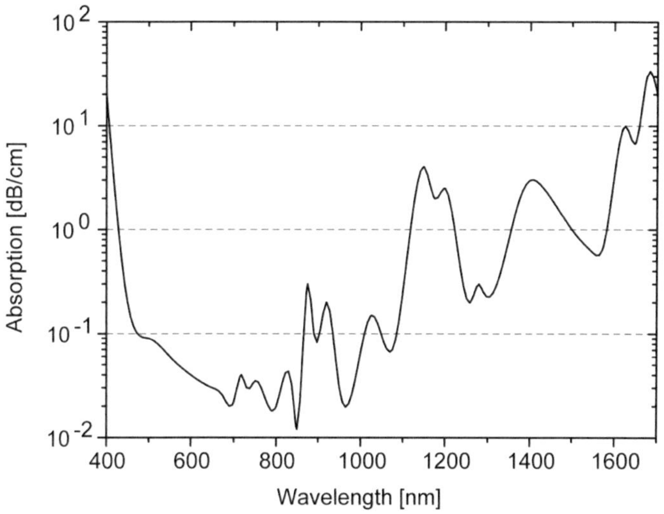

Fig. 3.17: Wavelength dependent absorption of ORMOCER-I from [65]. The material provides high transparency at 1550 nm wavelength and is therefore well-suited for optical interconnect waveguides.

The advantage of ORMOCER® mainly lies in the tailored material properties that are reached by chemical modification of the inorganic-organic hybrid material on the molecular scale. The ORMOCER® molecules consist of an inorganic poly-siloxane network that is generated by poly-condensation reaction and provides hardness and thermal stability of the material. By replacing Si atoms partially by atoms of other elements like Ti, Zr or Al the hardness, density as well as the refractive index of the resist can be modified. Apart from the poly-siloxane backplane ORMOCER® provides a second cross-linking reaction between the organic network builders. Depending on the groups used, these organic network builders provide optically induced cross-linking and can be used to tailor elasticity and thermal expansion. Finally, network modifiers can be used to tune elasticity, wetting behavior as well as the absorption behavior. The refractive index of ORMOCER® -I amounts to $n = 1.58$ at a wavelength $\lambda = 1550$ nm and is therefore comparable to the SU-8 [38] material used in the fabrication of PWB in Section 3.1-3.3. The resist in the non-polymerized stage, however, is liquid and can be processed according to the processed described for liquid, radical resist material in Section 1.3.2.

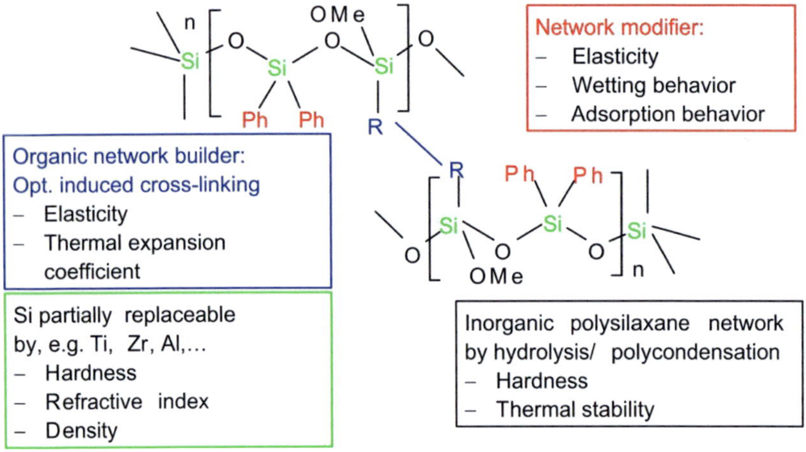

Fig. 3.18: Structure formula of ORMOCER-I [66]. The ORMOCER® molecules consist of an inorganic poly-siloxane network that is generated by poly-condensation reaction and provides hardness and thermal stability of the material. By replacing Si atoms partially atoms of other elements like Ti, Zr or Al the hardness, density as well as the refractive index of the resist can be modified. Apart from the poly-siloxane backplane ORMOCER® provides a second cross-linking reaction between the organic network builders. Depending on the groups used these organic network builders provide optically induced cross-linking, can be used to tailor elasticity or thermal expansion. Finally the network modifiers can be used to tune elasticity, wetting behavior as well as the absorption behavior.

The discussed photo resist is best structured at a lithography wavelength of $\lambda = 515$ nm. Since this lithography wavelength is not available at KIT facilities, the experimental setup at the Fraunhofer ISC Institute in Wuerzburg has been used for sample fabrication, see Fig. 3.19.

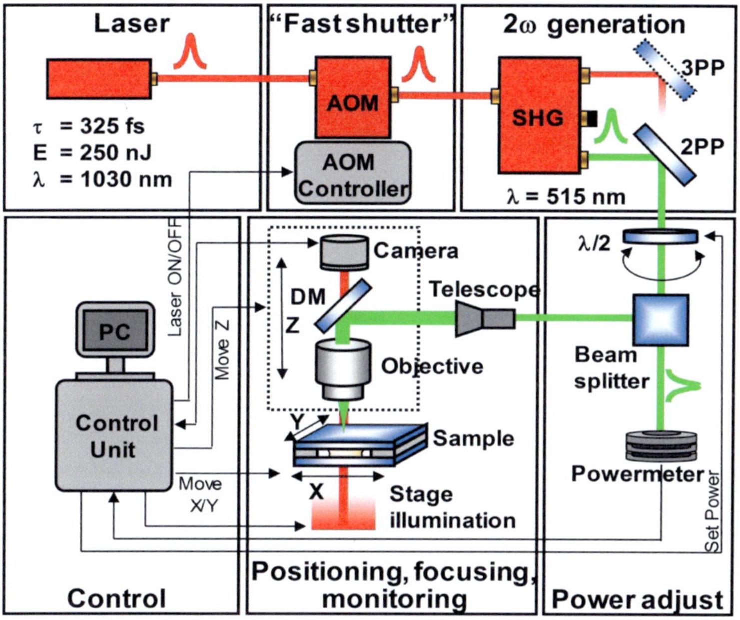

Fig. 3.19: Experimental lithography setup at Fraunhofer ISC Wuerzburg which has been used for the experiments performed in Section 3.4, figure adapted from [67]. The lithography wavelength of $\lambda = 515$ nm is generated from second harmonic generation of the input wavelength $\lambda = 1030$ nm. The system uses a repetition rate of 10 MHz and a pulse duration $\tau = 325$ fs. A movable $\lambda/2$ plate together with a polarization beam splitter is used to adjust the laser power used for lithography. Like in the piezo-based lithography system described in Section 1.3.3 this setup moves the sample with respect to the fixed objective position. A PC works as a control unit.

The lithography wavelength of $\lambda = 515$ nm in this setup is generated from second harmonic generation of the input wavelength of $\lambda = 1030$ nm. The system uses a repetition rate of 10 MHz and a pulse duration $\tau = 325$ fs. Like in the piezo-based lithography system described in Section 1.3.3, the sample is moved with respect to a fixed laser focus in this setup.

Photonic wire bonds with a round cross-section and a diameter of $d_{PWB} = 1.4$ µm are fabricated according to the helix writing strategy introduced in Section 2.4.2. The PWB trajectory is a straight line in this case. This trajectory was chosen due to the limitations of the non-standard experimental setup as well as to guarantee for comparability between the single PWB structures. When taking into account non-flat PWB structures on different chips the comparability of the structures is limited, since they will in general not encounter the same trajectory due to chip offsets and tilt.

The PWB waveguides in this experiment couple to nominally identical SOI nanowire waveguides. The SOI waveguides are down-tapered towards the end of the waveguide over a length of $l_{Si} = 60$ µm. Their tip width amounts to $w_{tip} = 80$ nm. The PWB waveguides are down-tapered towards the silicon waveguides to a tip width of $w_{PWB,taper,1} = 0.8$ µm. A SEM image of the fabricated PWB structures can be seen in Fig. 3.20.

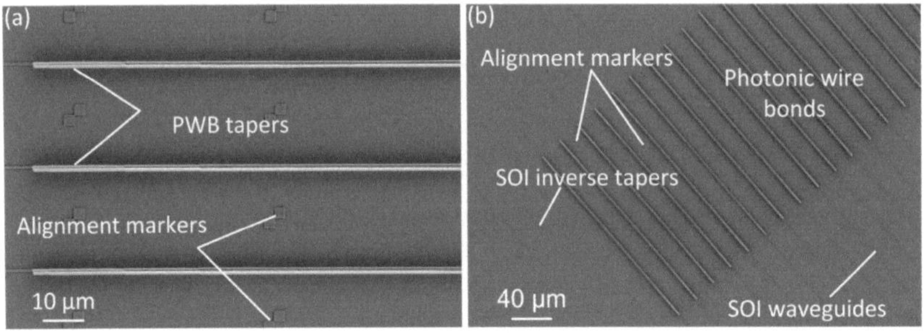

Fig. 3.20: SEM image of the fabricated ORMOCER-I PWB waveguides. (a) The photonic wire bonds have a circular cross-section with a diameter of $d_{PWB} = 1.4$ µm. The PWB waveguides are down-tapered towards the silicon waveguides to a tip width of $w_{PWB,taper,1} = 0.8$ µm. (b) The waveguide trajectory is a straight line in this case. The SOI waveguides are down-tapered towards the end of the waveguide over a length of $l_{Si} = 60$ µm. The PWB tip width amounts to $w_{tip} = 80$ nm.

The losses of five identical PWB waveguides on five identical SOI waveguides are compared in Fig. 3.21. The measurement is performed using a tunable laser source together with an optical spectrum analyzer, in a setup comparable to Fig. 3.12. Standard single-mode fibers in combination with conventional grating structures are used to couple light to and from the chip.

To obtain the insertion loss of the photonic wire bond, the spectrum of the PWB together with their SOI feed waveguides is first measured. The result is then normalized to the transmission spectrum of a same-length silicon reference waveguide on the same chip to isolate the contribution of the PWB. Average losses amount to 1.1 dB at a wavelength of 1550 nm with a standard deviation of 0.28 dB.

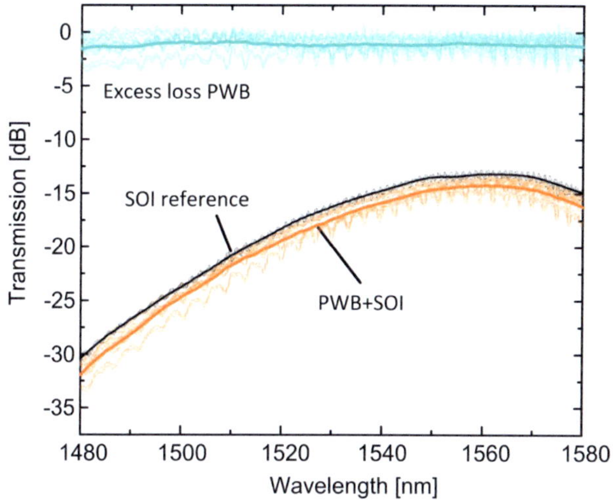

Fig. 3.21: Losses over five nominal identical flat on-chip PWB (blue): The measurement was performed using a tunable together with an optical spectrum analyzer. To obtain the insertion loss of the PWB, the spectrum of the PWB together with their SOI feed waveguides was first measured, see orange dotted lines. The result was then normalized to the transmission spectrum of a same-length silicon reference waveguide on the same chip (black line) to isolate the contribution of the PWB (blue dotted lines). The average loss of the five nominal identical PWB amounts to 1.1 dB at a wavelength of $\lambda = 1550$ nm, the standard deviation is 0.28* dB.

This experiment shows that the photonic wire bonding concept can be easily applied to different material platforms and lithography systems. The losses of the photonic wire bonds fabricated in this section lie within a reasonable tolerance for flat on-chip PWB structures.

4 Connections between Lasers and Silicon-on-Insulator Chips

After having intensively investigated the coupling to silicon-on-insulator (SOI) components in the last chapter, this chapter focuses on the interface between photonic wire bonds and a semiconductor laser, which is a key enabler for photonic multi-chip systems.

Due to the indirect bandgap of silicon, it is not possible to build a laser in the silicon photonics platform alone. The most prominent approaches to overcome this lack rely on wafer bonding or epitaxial growth of III-V lasers on top of the silicon platform [68], [69] or on Ge-on-Si co-integration [70]. Although these approaches have the potential of future large-scale integration, the device performance is still limited and lags behind the performance of conventional III-V laser sources. In addition to that, the mentioned examples all base on the use of high-cost, highest precision fabrication technology that might not be affordable for all applications.

The approach to overcome the laser issue in silicon photonics followed in this work, is to combine known-good laser dies – here the material platform of the laser die itself plays only a secondary role – with readily processed silicon photonic chips. Whereas this concept can also be followed by stacked-die packaging and flip-chip bonding [71], the presented approach using photonic wire bonding does not require high-precision mechanical alignment of photonic components with respect to each other.

While photonic wire bonding has been demonstrated to interconnect a silicon-on-insulator waveguide to a horizontal-cavity surface-emitting laser (HCSEL) [C4], this chapter deals with the coupling of photonic wire bonds to conventional edge emitting lasers.

The waveguide outlets in SOI devices can be directly accessed by 3D laser lithography and coupling to them can be managed *via* double-taper approaches or grating coupler. This is due to the fact that waveguide outlets are located on the top surface of photonic chips as described in the last chapters. In contrast to that, coupling to facet emitting devices is more challenging, since the writing beam cannot access the entire structure in an unperturbed way. One possible solution that has been investigated, is the use of an angled sub-mount that tilts the facet of the laser upward, so that a sufficient part of the lithography laser power can reach the facet of the laser and polymerize the resist. This option, although rather easy to implement, leads to uneven, non-standard packages that are undesired for commercial applications and the option was therefore not followed any further.

The second approach, which is discussed in detail in the following, is the direct coupling of photonic wire bonds to vertical device facets in a planar configuration. This option requires an elaborate control of the writing parameters, since the focus of the lithography tool is strongly disturbed in the proximity of the vertical device facet.

This chapter starts with the simulation of the optical interface between PWB and the laser source used in this experiment. Tapered sections as an integral part of the PWB are used to adapt the mode fields of the involved waveguides. The theoretical background, as well as the simulations on the correct writing parameters will be discussed in the following. After these more theoretical studies, the experimental verification of the PWB-based laser-to-chip coupling is given.

4.1 Spot-Size Conversion

The concept of photonic wire bonding offers a greatly simplified method to connect photonic components to modules compared to active alignment based techniques. The flexibility in the design of optical modules based on photonic wire bonds does, however, only play a decisive role, if the interconnect quality provided by the photonic wire bonds suffices. Due to the full three-dimensional capabilities of the lithography scheme the cross-section of the photonic wire bond can be freely chosen and varied along the trajectory of the bond. This offers a great flexibility to adapt the photonic wire bonds to different photonic components. Spot-size converters can be fabricated as an integral part of the photonic wire bond. However, since the mode fields of PWB and those of the waveguides to be interconnected are in general different from each other, interconnect losses due to this mismatch are indispensable as coupling to radiating modes cannot be completely avoided.

4.1.1 Mode Field Adaption

Mode field mismatch losses can be calculated from the overlap integral η between the complex optical field $\vec{E}_1(x,y,z')$, $\vec{H}_1(x,y,z')$ in the first waveguide and $\vec{E}_2(x,y,z')$, $\vec{H}_2(x,y,z')$ in the second waveguide [72]. The parameter z' denotes the coupling location.

$$\eta = \frac{\left| \frac{1}{4} \int_{-\infty}^{\infty}\int \left(\vec{E}_1 \times \vec{H}_2^* + \vec{E}_2^* \times \vec{H}_1 \right) \cdot \vec{e}_z \mathrm{d}x\, \mathrm{d}y \right|^2}{\frac{1}{2} \int_{-\infty}^{\infty}\int \mathrm{Re}\left\{ \vec{E}_1 \times \vec{H}_1^* \right\} \cdot \vec{e}_z \mathrm{d}x\, \mathrm{d}y \cdot \frac{1}{2} \int_{-\infty}^{\infty}\int \mathrm{Re}\left\{ \vec{E}_2 \times \vec{H}_2^* \right\} \cdot \vec{e}_z \mathrm{d}x\, \mathrm{d}y} . \quad (4.1)$$

The overlap on a dB scale is given by

$$\eta_{\mathrm{dB}} = 10 \log(\eta) \quad (4.2)$$

In general, an ideal interconnection between different photonic components is given, if the overlap integral of the devices specific mode fields is maximized. Mode fields can be matched either by changing the waveguide dimensions or the refractive index difference between core and cladding of one waveguide.

In photonic wire bonding the variation of waveguide dimensions is the preferred option to adapt the mode fields, whereas the refractive index of the PWB core and the surrounding cladding material are kept constant. This preference is caused by the fact that the involved refractive indices can only be influenced by the choice of the used resist or cladding material.

Apart from a good mode field overlap the chosen coupling structures have to encounter low loss in the transitions section between the interface cross-section and the PWB waveguide cross-sections. The coupling structures have to be tolerant against slight fabrication tolerances, and remain manufacturable in three-dimensional laser lithography.

4.1.2 Simulated Laser-to-Photonic Wire Bond Interface

The laser device studied in this section is an edge-emitting DFB laser [73] with a structure shown in Fig. 4.1 [73]. From the known refractive index distribution and geometrical dimensions, the cold cavity mode of the laser is simulated as can be seen from Fig. 4.1. The mode field diameter is different in *x*- and *y*- direction which already indicates that the ideal coupling between the laser and a photonic wire bond will require an elliptical or rectangular taper.

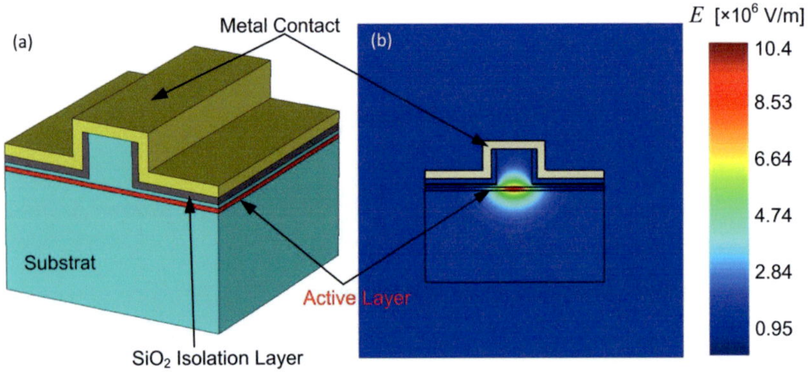

Fig. 4.1: Schematic drawing of the edge-emitting semiconductor laser: (a) perspective view: the active region is buried beneath a bulk material, an SiO_2 isolation layer as well as a gold contact pad. (b) simulated mode field of the cold cavity mode of the semiconductor laser, The energy in the mode is normalized to 1J [60].

The overlap integral between the simulated mode of the laser and the simulated mode of a photonic wire bond for different PWB width and height is calculated. Fig. 4.2 shows the overlap loss between the laser mode and the PWB mode. For an elliptical PWB ending cross-section, optimum coupling is reached for a horizontal diameter l_x of 2.7 µm and a vertical diameter l_y of 0.6 µm with an overlap loss as low as 0.23 dB. Additional taper-losses are negligible in this case – for a 40 µm long taper they amount to 0.04 dB in the case of optimum coupling – and are therefore not taken into account here. While the fabrication of photonic wire bonds with vertical diameters of 0.6 µm is possible in theory, the experimental realization of these structures will be challenging. The range of acceptable coupling loss for the simulated configuration, however, offers certain design flexibility, since larger vertical diameters of up to 2 µm lead to predicted losses below 0.5 dB.

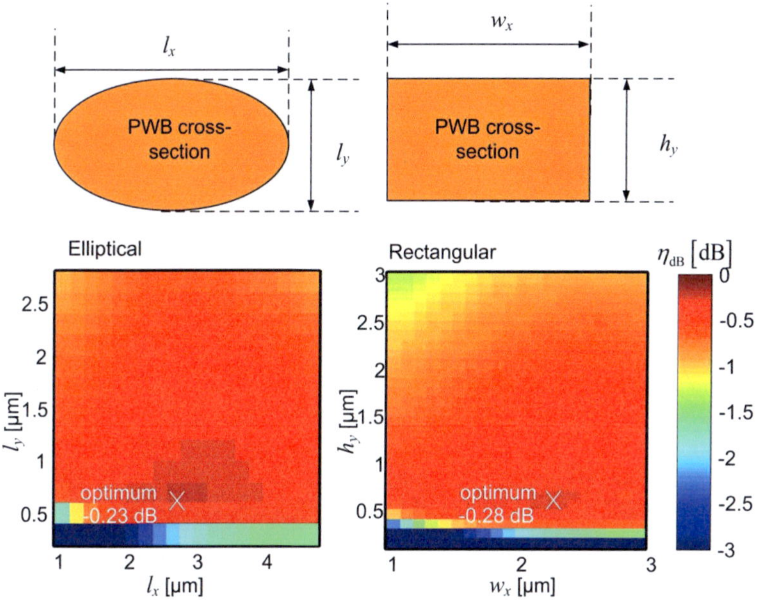

Fig. 4.2: Overlap loss between the simulated mode field of the laser (see Fig. 4.1) and the mode of a photonic wire bond with (a) elliptical cross-section; (b) resulting overlap integral between the simulated mode field of the laser and the mode of a photonic wire bond with a rectangular cross-section

For a rectangular PWB ending cross-section optimum coupling is reached for a width w_x of 2.2 µm and a height h_y of 0.6 µm with an overlap loss of 0.28 dB. In the case of an elliptical cross-section, the structure also allows for acceptable overlap losses below 0.5 dB with relaxed PWB dimensions of up to $2\,\mu m \times 2.5\,\mu m$. Comparable losses are also possible with the use of a rectangular taper cross-section, as can be seen in Fig. 4.2. b. The overall dimensions, however, are slightly smaller than in the case of an elliptical PWB cross-section.

4.2 Coupling to Vertical Device Facets

Coupling of photonic wire bonds to vertical device facets is a very challenging task. Due to the partial shadowing of the lithography beam – see Fig. 4.3 – undesired focus effects like intensity loss, the evolution of several intensity maxima, shift of focus as well as focus broadening can occur depending on the materials and geometries involved. In order to understand these processes in more detail and to be able to find the correct pre-distortion parameters the influence of a vertical device facet on the focus field distribution is simulated numerically. The work on this section has been performed in close collaboration with Maria Laura Gödecke [74].

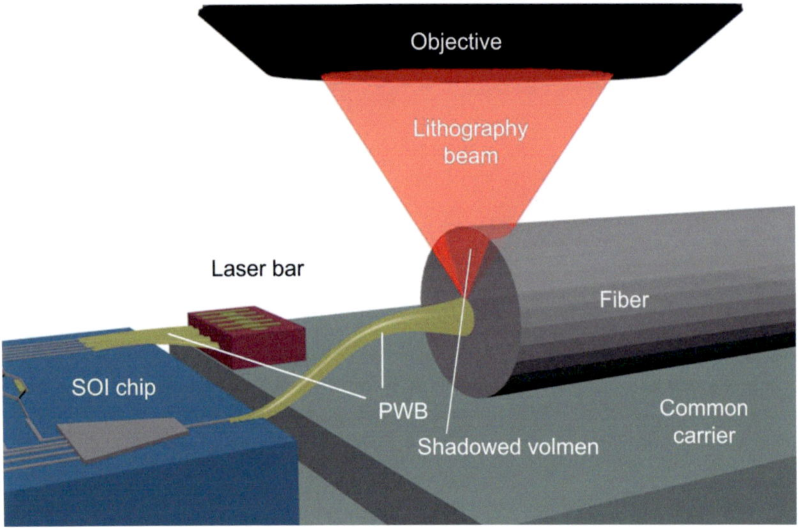

Fig. 4.3: Shadowing at device edges in the lithography system: Here the interconnection between laser devices, an SOI chip as well as an optical fiber by means of photonic wire bonds is displayed. All devices are pre-mounted on a common carrier with different height levels for rough pre-adjustment. The optical axis of all devices lies in a common plane. For the fabrication of planar multi-chip systems by photonic wire bonding, it is indispensable to couple to vertical device facets. The most crucial interfaces here are the facet of edge emitting lasers as well as the interface towards the standard single-mode fiber. The lithography beam is partially shadowed by the device edges and does not reach the desired focus position of the lithography system in an undisturbed way.

4.2.1 Numerical Simulation of Focal Intensity Distributions

As a starting point for numerical simulations a description of the focused laser beam in close proximity to the optical focus, i. e. in close proximity to the disturbing element is required. As high numerical aperture systems are used in the presented experiments, the involved optical fields cannot be described by the Gaussian beam approximation. The underlying assumption of the Gaussian beam approximation, i. e. the paraxial approximation is not valid in this case. As a consequence an expression for the electrical field distribution in close vicinity to the focus of the lithography system has to be derived.
In the following we will use the imaginary unit defined as $\mathrm{i}=\sqrt{-1}$. The propagation vector is $\vec{k}=\left(k_x,k_y,k_z\right)^{\mathrm{T}}$ with $\left|\vec{k}\right|=n\cdot w/c$ and $k_z=\sqrt{k^2-k_x^2-k_y^2}$. The angle Ω_1 is the solid angle under which the aperture area of the objective appears from the focal point P_1. The lithography objective is fully described by the matrix operator $\mathbf{M}$. The variable R represents the on-axis distance between the exit pupil of the objective and the focus. A schematic illustration of the aplanatic imaging system can be seen in Fig. 4.4 and a detail derivation of Eq. (4.3) is given in Appendix A.7.

With the matrix operator $\mathbf{M}$ describing the influence of the imaging system on the incident laser beam $\vec{E}_i$ an expression of the field in the focal region $\vec{E}(x,y,z)$ can be found following the theoretical work of Stamnes [75] and van de Nes [76], [77], to be

$$\vec{E}(\vec{r})=-\frac{\mathrm{i}R}{2\pi}\iint\limits_{\Omega_1}\frac{\mathbf{M}\tilde{\vec{E}}_i\left(k_x,k_y\right)}{\sqrt{k_z k}}e^{\mathrm{i}\vec{k}\vec{r}}\mathrm{d}k_x\,\mathrm{d}k_y. \tag{4.3}$$

The expression can be used to describe focusing of arbitrary incident laser beams $\vec{E}_i$ with its angular spectrum

$$\tilde{\vec{E}}_i\left(k_y,k_y\right)=\iint_{-\infty}^{+\infty}\vec{E}_i\left(x,y,0\right)e^{-\mathrm{i}\left(k_x x+k_y y\right)}\mathrm{d}x\,\mathrm{d}y. \tag{4.4}$$

For the evaluation of the double integral in (4.3) the corresponding fast Fourier transformation (FFT) algorithm [78] indicated by the symbol $\mathcal{F}_{k_x,k_y}$ is exploited.

$$\vec{E}(x,y,z) \propto \mathcal{F}^{-1}_{k_x,k_y} \left\{ \frac{\mathbf{M}\tilde{\vec{E}}_i(k_x,k_y)}{\sqrt{k_z k}} e^{ik_z z} \right\} \tag{4.5}$$

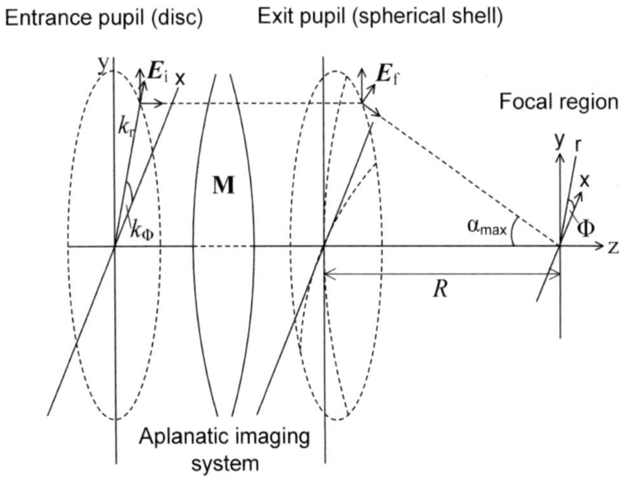

Fig. 4.4: Aplanatic imaging system adapted from [76]. Since an aberration-free imaging system is considered here Abbes sine condition holds, demanding that an incident light ray entering the system in a height *h* in the entrance pupil leaves the exit pupil in the same height *h*. The focusing can be described as a matrix operation between the entering and the exiting fields $\tilde{\vec{E}}_f(k_x,k_y) = \mathbf{M}\tilde{\vec{E}}_i(k_x,k_y)$.

This field distribution can now be used as a starting point for numerical simulation. Equation Eq. (4.5) is used to calculate the field distribution in a certain distance above the upper edge of the vertical device facet as can be seen in Fig. 4.5. This field distribution is loaded into an FDTD solver – here *Meep* [79] is used – and defined as a source for field propagation. A continuous light source with the center wavelength at λ=780 nm is used as it is in accordance with the used lithography system. For the simulations shown, the polarization of the incident field is set to right-handed circular polarization and the refractive index of the background material in the simulation box has a refractive index of $n = 1.52$, resembling the configuration in the lithography setup. For the sake of simplicity, the origin of the simulation box is chosen to coincide with the geometrical focus of the lithography system in all simulations. The device to be interconnected is included into the simulation domain defined by its geometrical dimensions and refractive index distributions. Propagating through the simulation domain, the field interacts with the predefined device. After the steady-state criterion has been met, the field distribution in the focus can be analyzed to identify the influence of the devices edges, facets and refractive index distributions.

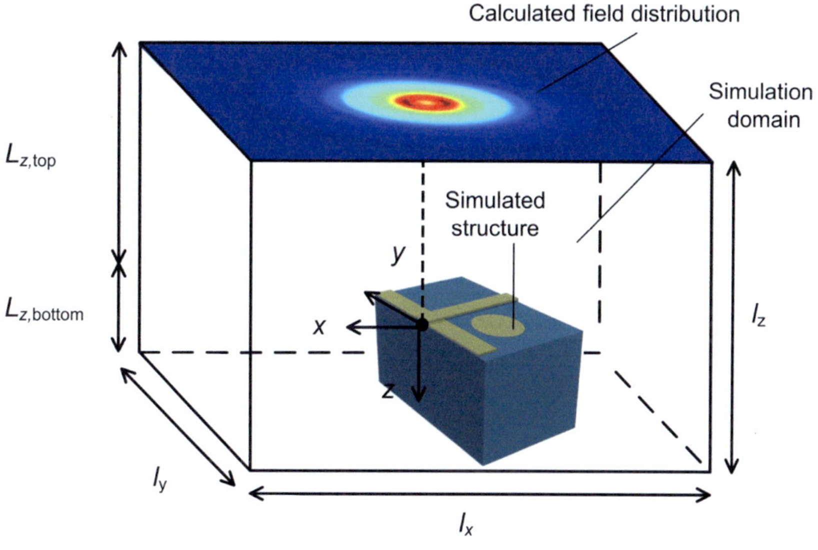

Fig. 4.5: The electrical field distribution is given in a certain height l_{Ztop} above the upper edge of the device that partially blocks the lithography beam. This field distribution serves as a source in FDTD simulation. The blocking device is located in the simulation box with l_x, l_y and l_z being the dimensions of the simulation box. This box has to be chosen big enough to allow for a convergent simulation.

In order to identify proper simulation parameters, such as resolution, domain size and boundary conditions, the focal field distribution in an unperturbed case, where the lithography beam can reach the focus without interacting with any devices in the simulation domain is simulated first. This simulation is compared to a calculated field distribution with the same starting conditions.

For the simulations presented in this section, the following parameters are chosen: $l_{z,top}=5$ µm, $l_{z,bottom}=1$ µm, $l_x=l_y=20$ µm. In the presented simulation the numerical aperture is fixed to NA = 1.3. The resolution Δx, which corresponds to the distance of sampling points on a cubic lattice in all three dimensions, is set to 25 nm. The factor S is the Courant factor that is set fixed to $S=0.5$ and c is the speed of light. The simulation time step Δt is defined by

$$\Delta t = \frac{S\Delta x}{c}. \tag{4.6}$$

Here the quantity $n_{\min}$ represents the lowest refractive index in the simulation domain and the factor dim denotes the dimensionality of the discussed problem. The stability criterion of $S < n_{\min} / \sqrt{\dim}$ is met in the discussed cases.

The semiconductor laser used in this section has already been introduced in Section 4.1.2. Fig. 4.6 shows a sketch of the device. A gold layer is covering the whole relevant sample surface. In the simulation this layer is modeled as a perfect reflector with a negative infinite permittivity. At both sides of the center rib a SiO_2 isolation layer follows. The lasers waveguide layer is buried inside the bulk material which has a refractive index of $n = 3.34$. Fig. 4.6. shows the position and dimension of the emitted light spot. Measured from the top edge the emission spot lies 2.633 µm under the device surface.

In the simulations, the center of the emission spot is positioned in the center of the simulation box, i. e. in the geometrical focus of the lithography system. By moving the device up and down along the axial direction, the influence of the distance between the upper edge and the focus on the focus quality can be simulated. By moving the laser further away or closer to the geometrical focus in the lateral direction, the behavior that will occur when the vertical facet is approached in direct laser writing can be studied.

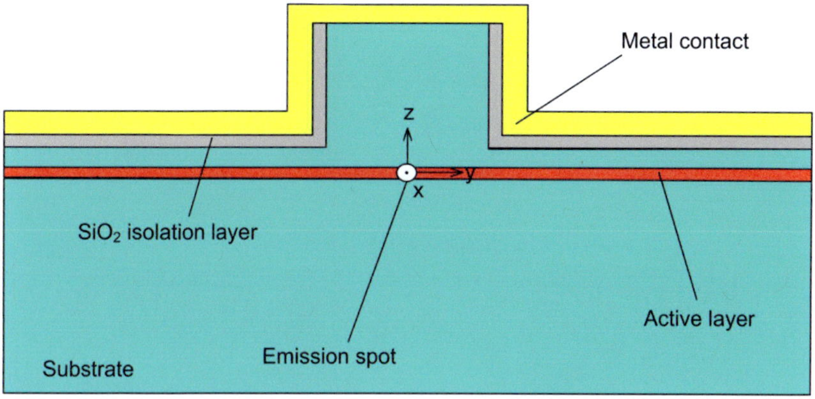

Fig. 4.6: Sketch of the used laser device and simulation coordinate system: The laser waveguide is buried beneath a bulk material, a SiO_2 isolation layer as well as a gold contact pad. In the simulation, the geometrical focus of the lithography system lies in the center of the waveguide core. The x-axis points out of the viewing plane. The y-axis is parallel to the epitaxial layers of the device, the z-axis is parallel to points towards the lithography objective.

Fig. 4.7 shows a series of simulations where the distance d_x between the geometrical focus and the vertical device axis is varied by moving the laser along the x-axis. In Fig. 4.7. d the distance between the geometrical focus and the laser facet is set to $d_x = 0.25$ µm so that the focus hits the laser facet. Displayed is a two-dimensional x-z-cut of the electrical field intensity $|E|^2$ plotted at the layer with highest $|E|^2$. The contour of the laser device is indicated by a white line. From Fig. 4.7. a to Fig. 4.7. d the distance between the laser facet and the geometrical focus is decreased from $d_x = 3$ µm to $d_x = 0.25$ µm. In a distance of $d_x = 3$ µm from the vertical device facet the focus field distribution resembles that of the unperturbed case, whereas the influence at smaller distances is clearly visible. The focus intensity distribution drops increasingly and the further symmetrical field distribution gets tilted. The tilt of the intensity distribution can be explained by the lack of the right-hand side field components. However, the overall focus shape still remains acceptable.

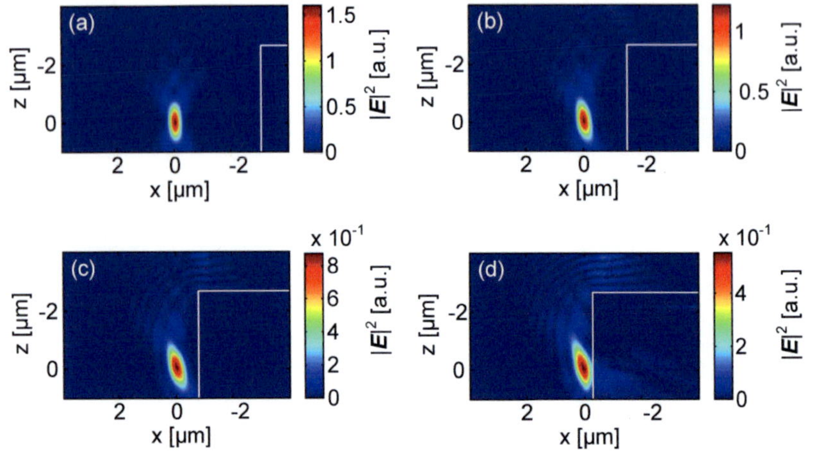

Fig. 4.7: *xz*-cuts of $|E|^2$ in the layer of highest $|E|^2$ showing the of the influence of the lateral distance d_x between the laser facet indicated by the white line and the geometrical focus of the lithography system: Displayed are the x-z-cuts of $|E|^2$ in the layer of highest $|E|^2$ at a distance of (a) $d_x = 3$ µm, (b) $d_x = 1.5$ µm, (c) $d_x = 0.75$ µm, and (d) $d_x = 0.25$ µm. With decreasing distance the focus intensity distribution tilts towards the left-hand side and its peak intensity drops.

The most severe influence on the focus quality that can be identified from the scan shown in Fig. 4.7 is the loss of focal intensity. Fig. 4.8 plots the normalized maximum value of the electrical field $|E|^2$ as a function of the distance between the geometrical focus and laser facet for different angles between laser facet and the optical axis, simulated at a resolution of 25 nm. In the practical case the vertical device facet might be tilted with respect to the optical axis of the lithography due to smallest perturbations in the pre-mounting of the laser source. To evaluate the influence of this tilt, the distance scan is performed under three different angles. At an angle of -5 the laser points upward, so that a smaller part of the lithography beam is disturbed by the upper edge of the device. At an angle of +5° the laser points downward. As it is clearly visible in the 0° curve, the focus intensity drops to about 23 % of its initial value, which means that it is necessary to increase the lithography laser power by a factor of 4.35 at the very laser facet compared to the unperturbed case, to deposit the necessary dose for two-photon polymerization. As expected an angle of -5°

results in a smaller focus intensity loss than in the case of a vertical device facet (0°) or even a downward pointing laser facet. From the simulated curves, power compensation curves can be derived and used in experiment.

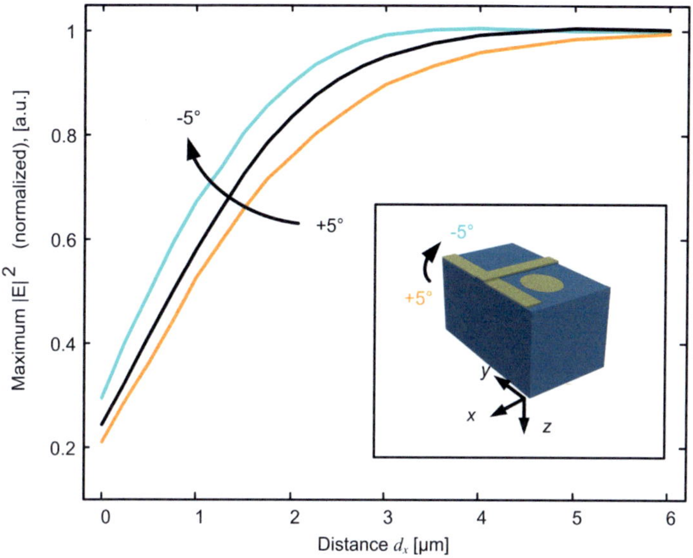

Fig. 4.8: Normalized maximum value of $|E|^2$ as a function of distance d_x between geometrical focus and laser facet along the x-axis for different angles between laser facet and the optical axis. At the laser facet the maximum intensity has dropped to 23 % of its initial value for the 0° simulation. At an angle of –5° the loss of focus intensity is smallest.

Apart from focal intensity loss the complete field distribution experiences a change when approaching the vertical device facet. To get a measure of the focal confinement, the second central moment of the electrical field distribution in all three dimensions is analyzed. The coordinates (x_{max}, y_{max}, z_{max}) represent the point of maximum field intensity. The second central moment in x-direction is given by

$$\mu_x = \frac{\iiint_{-\infty}^{\infty} (x - x_{max})^2 \left|\vec{E}(x,y,z)\right|^2 \mathrm{d}x\, \mathrm{d}y\, \mathrm{d}z}{\iiint_{-\infty}^{\infty} \left|\vec{E}(x,y,z)\right|^2 \mathrm{d}x\, \mathrm{d}y\, \mathrm{d}z} \tag{4.7}$$

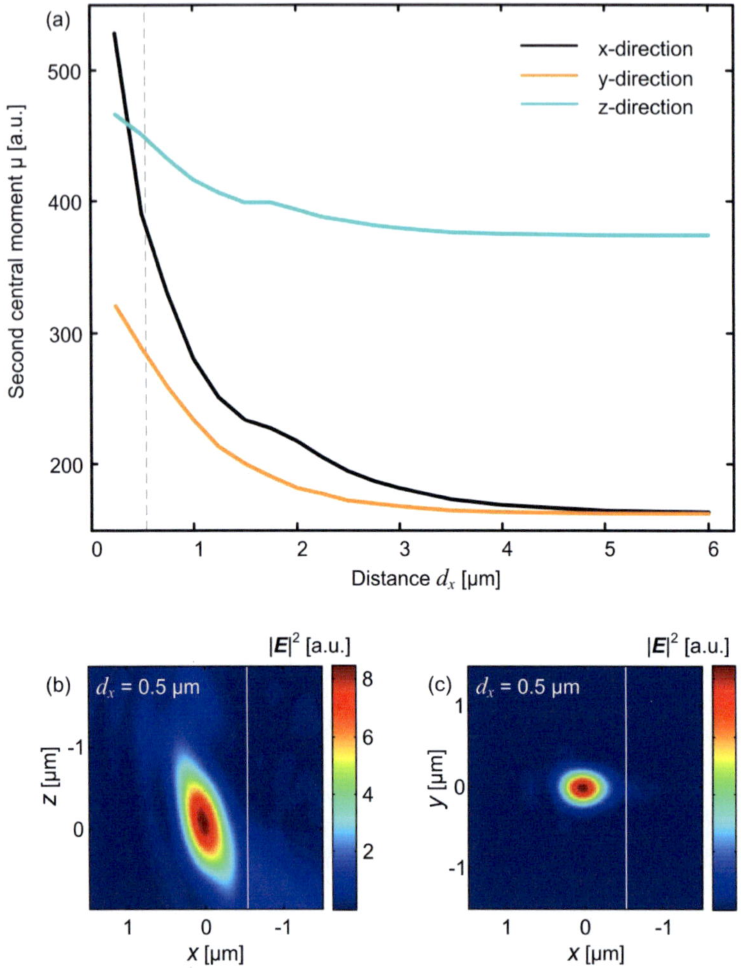

Fig. 4.9: Second central moment (a) for all three dimensions plotted versus distance d_x between vertical facet and geometrical focus. While the second central moment in y and z -direction only undergoes a slight change the increase of the second central moment in the x-direction is more pronounced, when the vertical device facet indicated by the white line in (b) and (c) is approached. However, the overall focus quality should still suffice for two-photon polymerization. (b) x-z cut and (c) x-y-cut through the point of maximum $|E|^2$ at a distance of $d = 0.5$ μm between the laser facet and the geometrical focus of the system.

Higher values of the second central moment represent a broad, smeared-out intensity distribution, whereas small values represent sharp foci. The second central moments for the *y*-, and *z*-component are defined analogously. The second central moment for all three dimensions versus distance d_x between vertical facet and geometrical focus is depicted in Fig. 4.9. While the second central moment in *y*- and *z*- direction only undergoes a slight change, the increase of the second central moment in the *x*-direction is more pronounced when the vertical device facet is approached. However, the overall focus quality should still suffice for two-photon polymerization.

To investigate the influence of the axial distance between the emission spot of the laser and the geometrical focus, an axial scan along the *z* direction is performed. As expected the metallization layer at the upper laser edge results in a strong reflection and interferences of the lithography beam in the vicinity of the upper edge. The focal intensity drops when the geometrical focus enters the shadow of the upper device edge, as shown in Fig. 4.10. The vertical dashed lines indicate the position the relevant range for photonic wire bonding.

The simulations indicate that the presence of the vertical laser facet together with its metallized upper edge results in a change of the focal intensity distribution. A compensation of the focal shape could be implemented but is not supposed to be required, whereas the loss of focal intensity has to be pre-compensated in order to deposit the necessary dose for two-photon lithography.

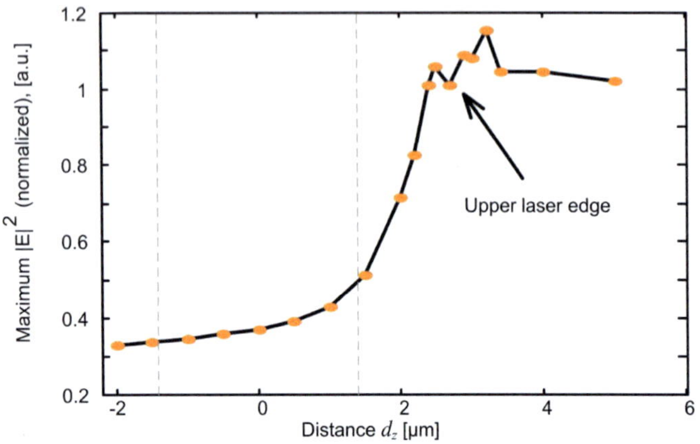

Fig. 4.10: Normalized maximum value of $|E|^2$ as a function of lateral distance d_z between the geometrical focus and the laser emission spot. The lateral distance d_x between the geometrical focus and the vertical device facet amounts to 0.3 μm in this case. The vertical dashed lines indicate the position of the emission spot and therefore the relevant range for photonic wire bonding.

4.2.2 Compensation of Focal Power Loss

Based on the theoretical and simulation findings, a laser-power compensation algorithm was implemented and integrated in the line writing strategy (see Section 2.4.2). Depending on the tilt of the device which is determined in the experiment, the corresponding curve from Fig. 4.8 is used as a compensation curve. For every writing line intersecting the relevant region, the line is subdivided into small segments and the distance to the vertical device facet is calculated for each of the segments. The quantity $\left|\vec{E}_{\text{undist}}\right|^2$ is the focus field intensity in the case where no disturbing device facet is present. Depending on the distance to the vertical device facet, the laser power is multiplied by a factor of $\left|\vec{E}_{\text{undist}}\right|^2 / \left|\vec{E}(x,y)\right|^2$ as a compensation factor.

4.2.3 Experimental Confirmation

To validate the simulation results, as well as the compensation algorithm, a first proof-of-principle experiment is performed. Test structures with the same shape and dimensions as the desired PWB-to-laser taper are fabricated at the backside of an edge-emitting semiconductor laser. The major ellipse axis was oriented parallel to the laser surface and was chosen to have a diameter of 3 µm, the minor axis of the ellipse was designed to be 1.5 µm both given at the interface towards the laser. The test structures had a length of approximately 50 µm, resembling the length of the taper used in later experiments.

The structures have been fabricated in line writing technique with a laser power increase according to the simulation results shown in Section 4.2.1. Fig 4.11 shows a SEM image of the resulting test structures. The structure in Fig. 4.11 a is positioned exactly in front of the laser emission spot. Fig. 4.11 b shows a close-up of the test structure at the interface between the central rib of the laser and the polymer taper. The structure dimensions are found to be in accordance with the design input and the adhesion between the test structure and the vertical laser facet is guaranteed although the interface is not perfectly smooth. One possible reason for this might be scattering of the lithography beams due to the rough surface of the gold top layer that was not included in the simulations.

This experiment shows that the simulation results together with the laser-implemented power compensation, based on the simulation results, can be used for structure generation at vertical device facets, which is a necessary requirement for the fabrication of photonic wire bonds at this interface.

Fig. 4.11: SEM picture of the fabricated test structure at the back side of an edge emitting semiconductor laser. The test structure has a length of approximately 50 µm and features elliptical cross-section geometry. The major ellipse axis was oriented parallel to the laser surface and was chosen to have a diameter of 3 µm, the minor axis of the ellipse was designed to be 1.5 µm both at the interface towards the laser. (a) The structure is positioned exactly in front of the laser emission spot. (b) Close-up of the interface between the polymer structure and the laser facet. The interface is not perfectly smooth which might be attributed to the influence of the rough surface of the gold top layer that was not included in the simulations. The rough material beneath the laser is an uneven layer of adhesive used for fixing the laser onto the metal substrate. The structure in the background is supposed to be a dirt particle that attached to the test structure during the development of the sample.

4.3 Fabrication

In order to obtain a suitable starting point for photonic wire bonding, the respective devices have to be pre-mounted in such a way that the distance between the interconnect regions, i. e. the emission spot of the laser and the SOI waveguide taper in this case, does not exceed a certain value. While distance tolerances in the direction of light propagation are relaxed ($\Delta x = 250\ \mu m \pm 15\ \mu m$ is not problematic) the lateral and vertical offsets between the devices are more stringent. In an ideal configuration relative positioning of the laser emission spot and the SOI taper tip is chosen, such that the laser emission spot is located $\Delta z = 20\ \mu m \pm 10\ \mu m$ above the SOI waveguide. The preferred horizontal distance in the direction perpendicular to the direction of light propagation is $\Delta y = 0\ \mu m \pm 50\ \mu m$. Since the SOI chip and the laser die have significantly different device thicknesses, the need for mounting on a common metal carrier with different height levels, arises, as can be seen in Fig. 4.12. The devices are manually positioned on the carrier and adhesive bonding is used for mounting of the devices. EPO-TEK® H20E from Epoxy Technology Inc., a two-component silver filled epoxy is used here [80]. This adhesive does not only fulfill the task of fixing the device to the carrier, but also provides a thermal and electrical contact between the bottom side electrode of the laser die and the metal carrier.

After a cleaning step, the sample is embedded into the negative-tone photo resist by drop casting of the liquid resist material. In this experiment *IP-Dip*™[39] was used.

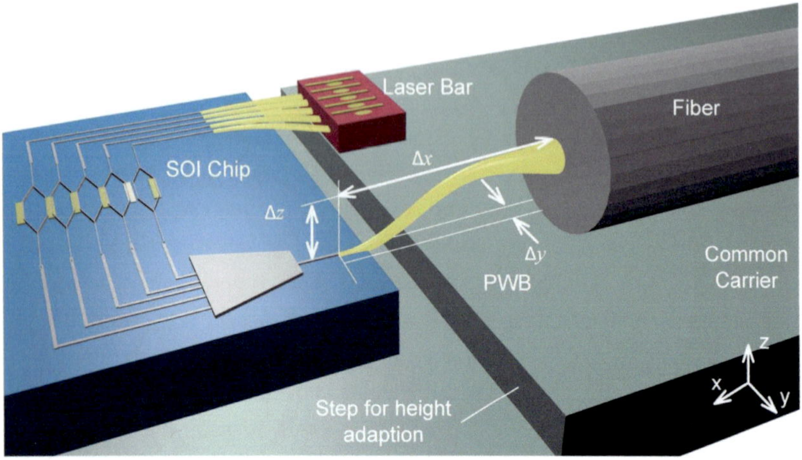

Fig. 4.12: Artistic illustration of the pre-mounting tolerances. The different components are mounted on a common carrier by means of adhesive bonding (adhesive not drawn here). The carrier features different height levels for rough adjustment of the different sample heights. In an ideal configuration relative positioning of the laser emission spot and the SOI taper tip is chosen, such that the laser emission spot is located Δz =20 µm±10 µm above the SOI waveguide. The preferred horizontal distances are smaller than Δx = 250 µm ±150 µm in the direction of light propagation and Δy =0 µm ± 50 µm perpendicular to the direction of light propagation. Those tolerances can easily be met with standard pick-and-place machinery.

The measurement of the interconnect position is performed manually in this experiment. The lateral positions of the inverse SOI taper as well as the position and direction of the central rib of the laser are visually determined from the wide-field image taken by the microscope camera through the lithography objective. The vertical position is adjusted manually by optimizing the focus of the image. From the position and direction of the lasers central rib as well as one additional point on the chip surface the position and direction of the emission spot are calculated, see Fig. 4.13.

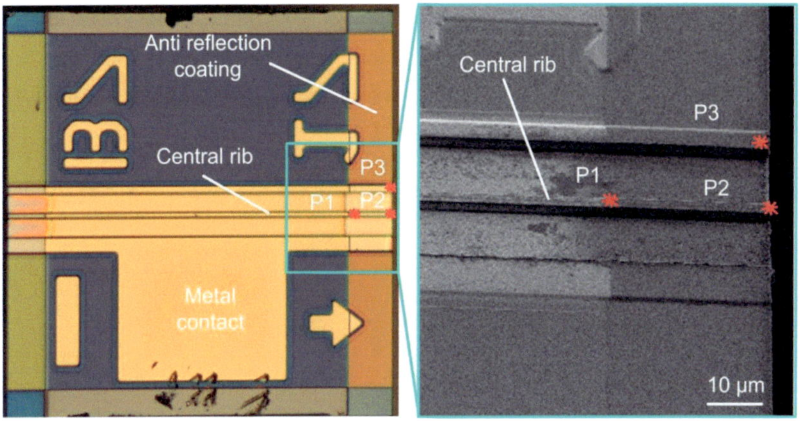

Fig. 4.13: Illustration of the port detection procedure for the edge-emitting semiconductor laser. Two points (P1 and P2) are measured on the central rib of the laser. From these two points the tilt of the device in the *x*-*y*-plane can be obtained. A third point (P3) on the surface of the laser (same level as the surface of the central rib) is measured for the calculation of the rotation of the device. Knowing the device geometry the measurement of these three points is sufficient to calculate the position and direction of the laser emission spot.

The trajectory of all PWB discussed in this section are calculated according to the procedure described in Section 2.2. The PWB are fabricated in line writing technology as described in Section 2.4.2. The distance between single writing lines is set to 100 nm in both horizontal and axial direction. In order to compensate for the focal power loss introduced by the lasers vertical device facet, the laser power compensation as described in Section 4.2.2 is exploited and in each point in the vicinity of the vertical device facet the corresponding laser power is applied. The PWB structuring is performed in the galvanometer-mirror based lithography setup featuring an immersion objective (40× magnification) with a large numerical aperture (NA = 1.3). During the lithography step, the objective is in direct contact with the resist material in a configuration as depicted in Fig. 1.4.

After the lithography step, the sample is developed in PGMEA (**p**ropylene **g**lycol **me**thyl **e**ther **a**cetate, 1-methoxy-2-propanol acetate) for seven minutes followed by a 2-minute bath in isopropyl alcohol. An additional drying step in a critical point dryer is applied to the samples depicted in Fig. 4.14 and Fig

4.15. This step serves to minimize capillary forces that might occur in the gap between the SOI and the laser chip during drying. In Fig. 4.14, Fig. 4.15 and Fig. 4.16 SEM images of three fabricated laser-to-chip PWB devices are shown.

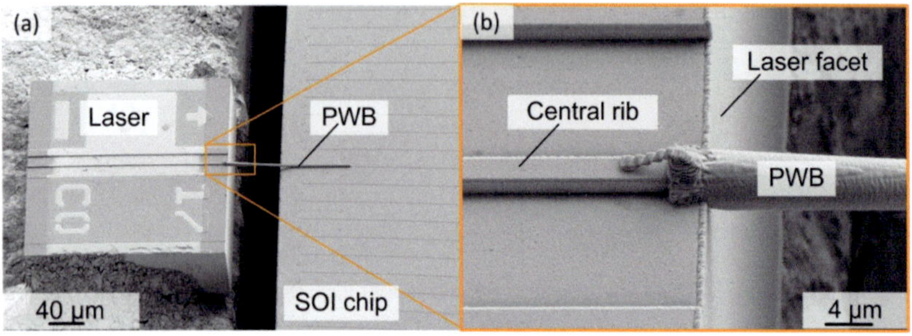

Fig. 4.14: SEM image of a fabricated laser-to-SOI PWB prototype. (a) overview picture. The photonic wire bond features a relatively straight trajectory. It bridges a gap of 102 µm and a vertical offset of 18 µm. The PWB taper cross-section at the laser interface was designed to be elliptical with a major axis of 5 µm and a minor axis of 3.7 µm. The close-up of the coupling region (b) shows that due to fabrication tolerances the PWB is located too high with a comparison to the designated z-position which will result in enhanced optical losses of the structure.

The PWB structure shown in Fig. 4.14 bridges a gap of 102 µm and a vertical offset of 18 µm. The PWB taper towards the laser facet has a length of 40 µm, its cross-section is designed to be elliptical with a major axis of 5 µm and a minor axis of 3.7 µm both measured at the laser interface. These dimensions are rather big compared to the optimum coupling configuration obtained from simulations in Section 4.1.2. They are chosen to ensure good adhesion at the laser facet. The PWB waveguide section features an elliptical cross-section with a major axis of 2.8 µm and a minor axis of 2 µm. The taper towards the silicon waveguide has a rectangular cross-section with a tip width and height of $w_{\mathrm{PWB,taper,1}} = 0.76$ µm by $h_{\mathrm{PWB,taper,1}} = 0.45$ µm. The length of the PWB taper structure towards the SOI waveguide is $l_{\mathrm{PWB,taper}} = 60$ µm. The SOI waveguide in the un-tapered region has a width of $w_{\mathrm{Si}} = 500$ nm and a height $h_{\mathrm{Si}} = 220$ nm. The tip width of the inverse SOI taper amounts to $w_{\mathrm{tip}} = 80$ nm and the length of the SOI taper amounts to $l_{\mathrm{Si}} = 60$ µm.

The close-up of the coupling region Fig. 4.14. b shows that due to fabrication tolerances the PWB is located too high in comparison to the designated z-position, which results in enhanced optical losses of the structure.

In Fig. 4.15, the result of the repeated experiment with a corrected height adjustment is shown. The PWB structure now bridges a gap of 110 µm and a vertical offset of 38 µm. The other parameters are comparable to the structure in Fig. 4.14. Due to the longer PWB structure a more pronounced (absolute) shrinkage of the resist in the drying step after development is observed. This effect causes a small gap between the laser facet and the PWB taper as can be seen in Fig. 4.15. b.

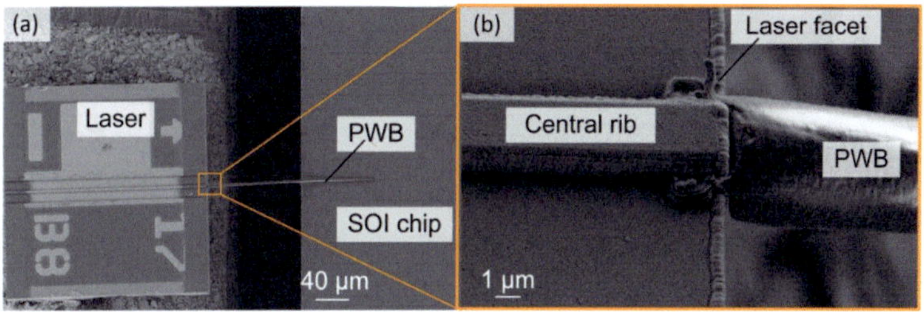

Fig. 4.15: Repeated experiment with corrected height adjustment: (a) overview picture, the PWB structure bridges a larger gap of 110 µm and a more pronounced vertical offset of 38 µm. All other parameters are comparable to the structure in Fig. 4.14. (b) due to shrinkage of the resist during drying after the developing step and caused by a presumably too low laser power at the facet the PWB does not attach directly to the facet but a small gap is visible between the PWB and the laser. Small side-arms hold the PWB in place.

Due to the strong shrinkage predominantly observed in the drying step after development, several follow-up experiments resulted in a complete detachment of the PWB from the laser facet as well as of deformation of the trajectory. Measures to overcome the shrinkage issue, like the artificial elongation of the trajectory as a compensation structure, as well as the use of additional anchoring structures at the laser interface did not lead to the desired result.

As the resist shrinkage is observed to happen in the drying step after the development, the PWB was immersed into index matching oil directly after development without drying it in a follow-up experiment. An optical image of the resulting device can be seen in Fig. 4.16. The photonic wire bond depicted here spans a length of 110 µm and a vertical offset as low as 1 µm. In order to pre-compensate for resist shrinkage, the trajectory was enlarged by adding an apex of approximately 10 µm in the middle of the trajectory.

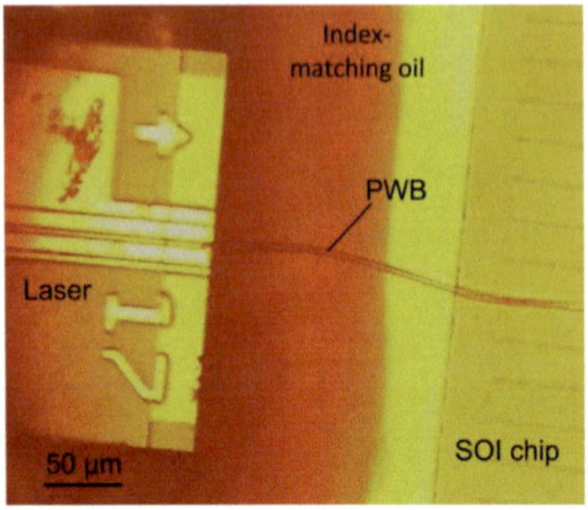

Fig. 4.16: Optical image of a PWB laser-to-chip connection immersed in index matching liquid. The structure has been immersed in the oil directly after the development step without drying the sample. The PWB structure is attached to the laser and the trajectory did not change during the process. The contour line of the PWB close to the laser facet is traced for better visibility.

The PWB taper towards the laser facet has a length of 40 µm, its cross-section is designed to be elliptical with a major axis of 3.2 µm and a minor axis of 1.7 µm both at the laser facet, to be closer to the simulated optimum. The PWB waveguide section features an elliptical cross-section with a major axis of 2.8 µm and a minor axis of 2 µm. The taper towards the silicon waveguide has a rectangular cross-section with a tip width and height of $w_{\mathrm{PWB,taper,1}} = 0.76$ µm by $h_{\mathrm{PWB,taper,1}} = 0.45$ µm. The length of the PWB taper structure towards the SOI waveguide is $l_{\mathrm{PWB,taper}} = 60$ µm. The SOI waveguide in the un-tapered region has a width of $w_{\mathrm{Si}} = 500$ nm and a height $h_{\mathrm{Si}} = 220$ nm. The tip width of the inverse SOI taper amounts to $w_{\mathrm{tip}} = 80$ nm and the length of the SOI taper amounts to $\mathrm{l}_{\mathrm{Si}} = 60$ µm. Alike the structures in Fig. 4.14 and Fig. 4.15, the PWB waveguide is fabricated from *IP-Dip*™[39] in the line writing technique described in Section 2.4.2 using the galvanometer mirror based lithography setup. The laser power towards the laser facet has been increased according to the values obtained from simulations, see Fig. 4.7.

After the development step in PGMEA, the sample is directly immersed in index oil to avoid a drying of the structure. Fig. 4.16 shows an optical image of the sample immersed in index oil. In a potential future step the immersion in oil can be replaced by the immersion in a dedicated polymer cladding material. The PWB structure is attached to the laser facet and the PWB trajectory is in accordance with the design.

4.4 Characterization

This section deals with the characterization of the samples described in the previous paragraph. For the measurement of the PWB excess loss a reference for the laser output power is required. This output power is measured versus the laser current by coupling a lensed fiber to the laser under discussion prior to mounting of the SOI chip and prior to the fabrication of the PWB waveguide. Measuring the output power directly by an integrating sphere would be preferred, but is prevented by the device/substrate geometry; the entrance of the integrating sphere cannot be brought close enough to the emission spot of the laser to capture the entire divergent laser beam.

To investigate the losses between lensed fiber and laser, the lensed fiber measurement of a reference laser that has been mounted on a different carrier is compared to the direct measurement of the output power of the same laser by an integrating sphere. The additional losses introduced by the lensed fiber coupling amount to 2.17 dB. These losses are essentially constant above lasing threshold. The power lost due to lensed fiber coupling is taken into account in the reference measurement of the laser used in the experiment.

For the measurement of the PWB loss the setup depicted in Fig. 4.17 is used. The laser is connected to a current source. Light from the laser is then transmitted through the photonic wire bond into an SOI waveguide.

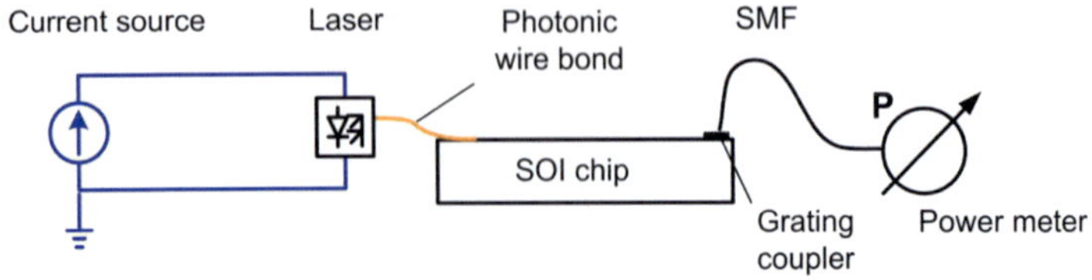

Fig. 4.17: Schematic illustration of the characterization setup: The laser is connected to a current source. Light emitted by the laser is guided through the PWB and the SOI. At the exit of the SOI waveguide light is coupled to an SMF by grating couplers and measured in a power meter.

The SOI waveguide is coupled to an SMF by means of a standard grating coupler [3]. The optical output power is then measured by a power meter.

The optical excess losses of the PWB have been obtained from a measurement of the whole assembly consisting of the laser, the PWB, the SOI waveguide, a grating coupler and an SMF. This measured output power has then been normalized to the pre-characterized optical output power of the laser. A transmission measurement of a single SOI waveguide coupled to two SMF by means of two grating couplers is performed to identify the loss contribution of the SOI and the grating coupler. The total losses measured in the whole assembly have than been divided by the losses in this reference measurement (divided by two, since only one grating coupler is involved in the discussed setup). The measured mean excess loss of the PWB structure displayed in Fig. 4.14 are displayed in Fig. 4.18 and amount to 8.87 dB. These rather high insertion losses are attributed to a vertical misalignment of the PWB with respect to the laser emission spot.

The optical output power and losses of the device immersed in oil directly after development, which is shown in Fig. 4.15, are displayed in Fig. 4.19. The measurement has been obtained in the same process as described for the previous sample. The mean excess loss of the PWB amounts to 8.44 dB in this case. The measurement, however, has been disturbed by residuals of undeveloped resist on the SOI waveguide. The measurement shows that the immersion of the sample right after the development in principle works, but has to be further improved in order to obtain competitive insertion losses.

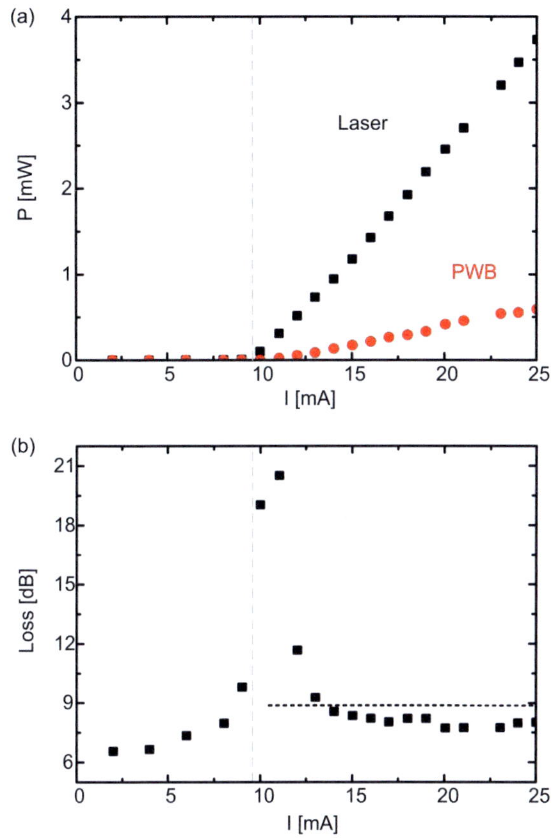

Fig. 4.18: Measured output power and optical loss of the device displayed in Fig. 4.14 (a) The optical output power versus current of the laser (black squares) has been obtained from a reference measurement before the fabrication of the PWB. The measurement has been obtained by lensed fiber measurement. The power losses of the lensed fiber have been added. The output power of the PWB (red circles) is measured through the whole assembly of PWB, SOI waveguide, grating coupler and single mode fiber. To obtain the power coupled to the SOI waveguide, the power lost in the SOI waveguide and the grating coupler is added up here. (b) Loss of the PWB connecting laser and SOI waveguide versus. Above the laser threshold, the mean excess losses of the PWB amount to 8.87 dB.

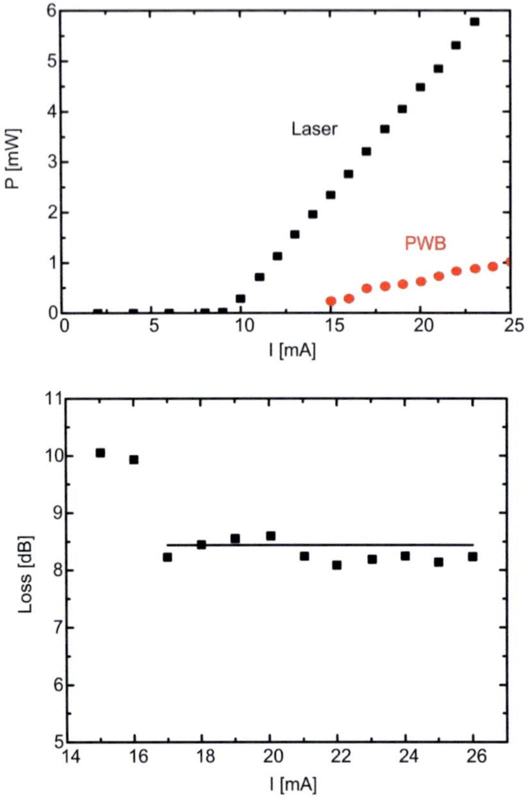

Fig. 4.19: Measured output power and optical loss of the device displayed in Fig. 4.16.(a) The optical output power versus current of the laser (black squares) has been obtained from a reference measurement before the fabrication of the PWB. The measurement has been obtained by lensed fiber measurement. The excess losses of the lensed fiber have been added. The output power of the PWB (red circles) is measured through the whole assembly of PWB, SOI waveguide, grating coupler and single mode fiber. To obtain the power coupled to the SOI waveguide, the power lost in the SOI waveguide and the grating coupler is added up here. Below a current value of 15 mA no further measurement points could be obtained due to a destruction of the sample. (b) excess loss of the PWB connecting laser and SOI waveguide versus current. Above the laser threshold the mean excess losses of the structure amount to 8.44 dB. The measured values at input currents of 15 mA and 14 mA have not been taken into account for the average optical losses since the device showed tremendous power fluctuations during measurement.

5 Connections between Fibers and Silicon-on-Insulator Chips

In each multi-chip system, information has to be transferred to the outside world sooner or later. In the case of optical communications this means that at a certain stage it becomes indispensable to connect devices to an optical (single mode) fiber. Optical fibers offer huge transmission bandwidth at a relatively low optical loss, 0.2 dB/km is a typical value. Low-loss and broadband fiber-to-chip coupling is thus one of the major tasks in optical input/output technology.

Out-of-plane coupling using grating couplers is considered one of the prime candidates in silicon photonics. These couplers that have already been introduced in Chapter 1, use diffraction gratings etched into the top surface of silicon-on-insulator nanowire waveguides [3]–[10]. When properly designed, such structures can simultaneously act as couplers and integrated polarization splitters [4]. Coupling efficiency can be improved by using bottom mirrors [9], [10], silicon overlays [5] or numerically optimized grating designs [6], leading to insertion losses between 1 dB and 2 dB for advanced devices. Grating couplers enable optical access anywhere on the chip surface. However, 1 dB-bandwidths are usually limited to typical values between 40 nm and 50 nm and the coupling scheme makes an out-of-plane scattering indispensable and therefore requires nonstandard packages.

Coupling to SOI nanowire waveguides through grating couplers relies on the highly precise positioning of fibers with respect to the grating couplers. The coupling process is mainly based on intricate active alignment techniques, where the coupling efficiency is dynamically monitored and optimized, and often involves manual fabrication steps. The use of photonic wire bonding to overcome these difficulties will be discussed in this chapter.

Similar to the case of laser-to-SOI coupling, the coupling of photonic wire bonds with a fiber requires structure generation at vertical device facets, if the fiber axis is oriented in the same plane as the SOI waveguides. The presence of the device facet, however, will introduce beam deformation and focal intensity loss. One option to reduce beam blockage is the use of fibers with tilted facets. However, the use of angled fibers has been decided against since the use of angled fibers requires a second adjustment step to align the fiber with respect to the lithography system. This limits the potential for future industrial applicability of the process and is therefore not desirable. Moreover, the fabrication of angled fibers with traditional equipment, like standard diamond cleavers often results in unpredictable cleaving and high surface roughness as can be seen in Fig. 5.1. Our focus therefore lies on coupling PWB to flat cleaved optical fibers, in a configuration, where the fiber axis is oriented perpendicular to the lithography systems optical axis, i. e. parallel to the SOI waveguide. The following section includes the simulation of the optical interface between fiber and photonic wire bond. The second section of this chapter will discuss the simulation of the lithography systems focal intensity in the vicinity of the fiber facet. In the last paragraph, the experimental results on fiber-to-fiber as well as fiber-to-chip coupling are provided. The work on the coupling between a four-core fiber and four SOI waveguides is discussed in Chapter 5.7.

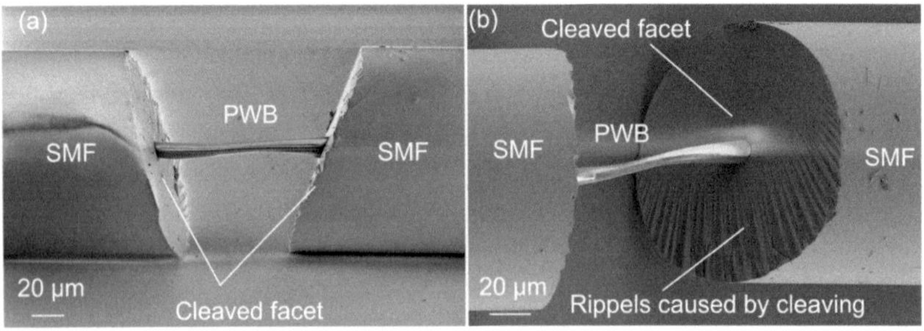

Fig. 5.1: Photonic wire bond between two standard single mode fibers cleaved at an angle of approximately 10°. (a) The use of cleaved fibers can reduce the beam blockage that is caused by the vertical fiber facet as well as the upper edge of the fiber. (b) Due to the cleaving process ripples and cracks can occur that hinder the optical transmission.

5.1 Mode Field Adaption

The mode fields of standard single mode fibers like *Corning® SMF*-28 [81] and a PWB have very different dimensions. The typical mode field of an SMF amounts to 10.4 µm whereas the mode field diameter of a single mode PWB is in the range of 2 - 2.5 µm. In order to adapt the mode fields, an up-tapered section as part of the photonic wire bond is used, as it is displayed in Fig. 5.2. In the case of a taper with a circular cross-section, the taper starts with the PWB waveguide diameter $d_{\mathrm{PWB}} = d_{\mathrm{taper,1}}$ over a length $l_{\mathrm{up\text{-}taper}}$ the diameter is linearly increased to a final diameter $d_{\mathrm{taper,2}}$.

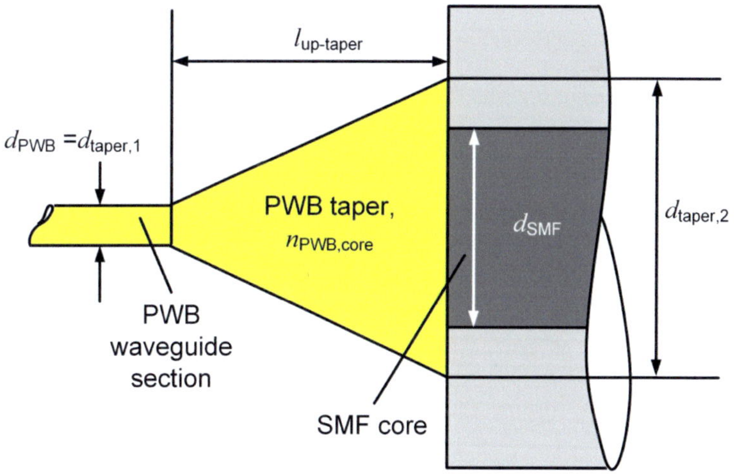

Fig. 5.2: Structure of the linear up-taper used at the interface between the SMF and the PWB. The taper starts with the circular cross-section of the PWB ($d_{\mathrm{PWB}} = d_{\mathrm{taper,1}} = 2$ µm) and ends with a circular cross-section of larger diameter $d_{\mathrm{taper,2}}$ at the fiber end face.

The overall loss of this taper is calculated as the sum of the overlap loss between taper and fiber, and the propagation loss in the taper. The overlap losses are calculated between the fundamental eigenmode of the PWB with a specific taper ending cross-section and the fundamental eigenmode of the standard single mode fiber, see Eq. (4.1). To calculate the transmission losses of the tapered section, the transmission between the fundamental eigenmode of the elliptical PWB in the waveguide section and the fundamental eigenmode of the PWB with the specific taper ending cross-section is simulated in a finite integration technique solver (FIT solver).

In all the simulation a PWB refractive index $n_{\mathrm{PWB,core}} = 1.53$ is applied, which corresponds to the refractive index of IP-Dip^{TM}. The background refractive index is set fixed to $n_{\mathrm{PBW,clad}} = 1.34$, to resemble the refractive index of the desired background material *Cytop*^{TM} [59]. The simulated overlap mismatch loss is indicated as red crosses in Fig. 5.3, whereas the transmission losses for different taper diameters are indicated by blue crosses. The overall insertion loss of the taper section is given by the black crosses. Here, a linear up-taper with circular cross-section is used. As can be seen from Fig. 5.3 the optimum coupling diameter depends on the length of the chosen taper.

While at small taper lengths the transmission related losses play a dominating role, their influence gets lower with higher taper length and the insertion loss of the structure asymptotically reaches the minimum loss dictated by the overlap-mismatch loss.

In the experimental case the taper length might be limited due to practical reasons such as small device distance or high lateral and horizontal offset. Therefore the optimum taper diameter has to be chosen depending on the conditions given by the experimental setup, keeping in mind that longer taper structures in general lead to lower insertion losses for this structure.

The simulation results depicted in Fig. 5.3 have been obtained from the simulation of linear tapers with a circular cross-section. In contrast to this data, the simulation results shown in Fig. 5.4 are obtained from independently varying *x*- and *y*-radii of an elliptical taper. Since the PWB waveguide section also has an elliptical cross-section (2.8 µm × 2 µm in this setup), these simulations showed that the overall minimum is even a bit lower than the minima indicated by Fig. 5.3.

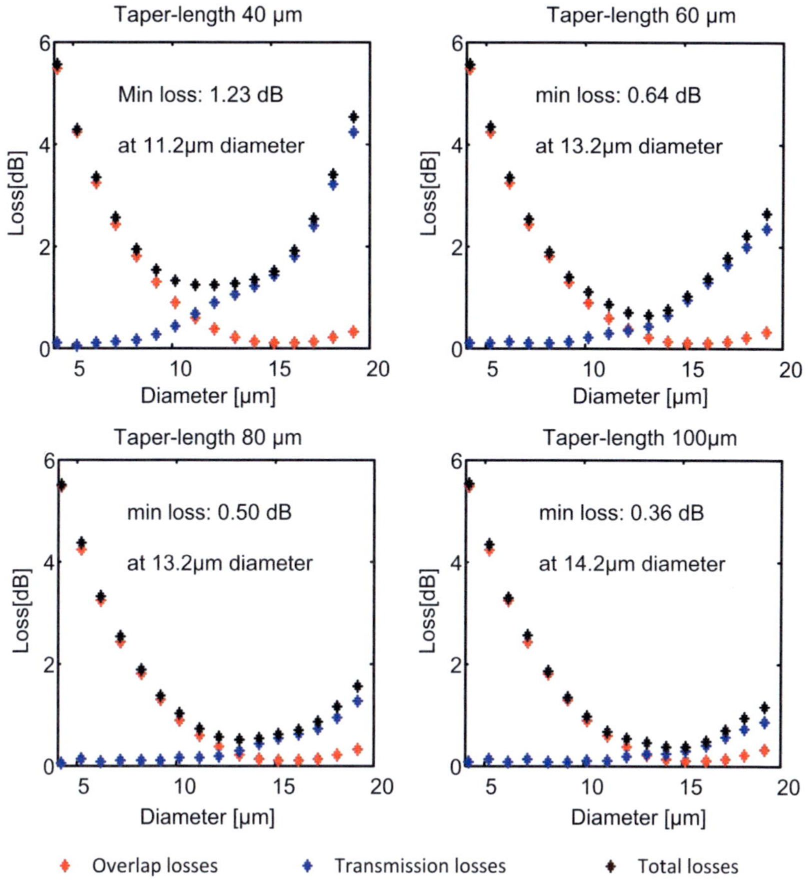

Fig. 5.3: Simulated insertion loss vs. taper diameter of an up-tapered PWB-to-fiber interface, as it is depicted in Fig. 5.2 for four different taper length. Here, the dimensions of the elliptical PWB waveguide section amount to 2.8 µm × 2 µm. In all the simulation a PWB refractive index $n_{\mathrm{PWB,core}} = 1.53$ is applied. The background refractive index is set fixed to $n_{\mathrm{PBW,clad}} = 1.34$. The simulated overlap mismatch loss is indicated as red crosses, the transmission losses for different taper diameters are indicated by blue crosses. The overall insertion loss of the taper section is given by the black crosses. In this simulation a linear taper with circular cross-section is used. The optimum coupling diameter depends on the length of the chosen taper. While at small taper length the transmission related losses play a dominating role, their influence gets lower with higher taper length and the insertion loss of the structure is asymptotically reaching the minimum loss dictated by the overlap-mismatch loss.

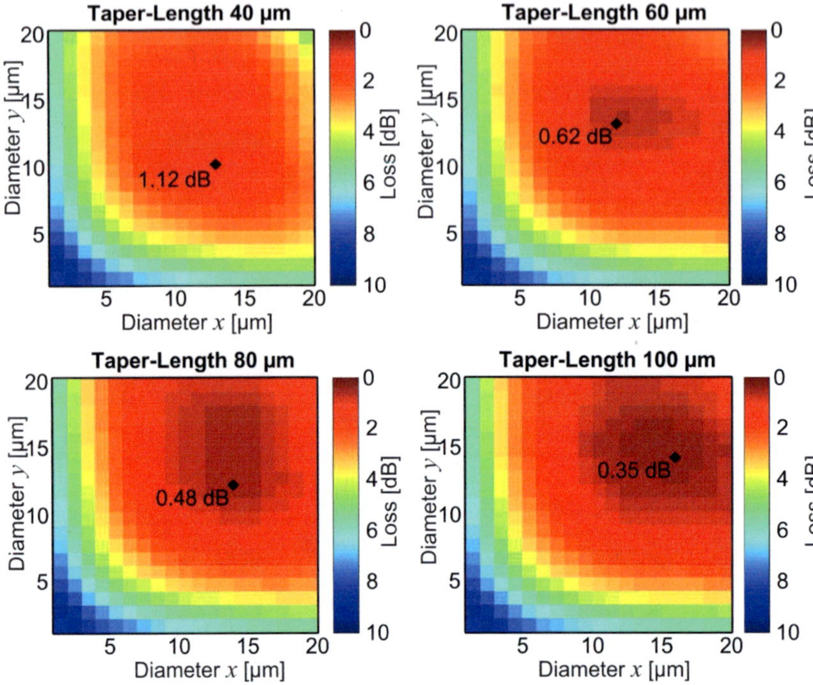

Fig. 5.4: Simulated insertion loss vs. taper diameter of an up-tapered PWB-to-fiber interface, for four different taper length. Here, an elliptical taper cross-section is applied. The dimensions of the elliptical PWB waveguide section amount to 2.8 µm × 2 µm. In all the simulation a PWB refractive index $n_{\mathrm{PWB,core}} = 1.53$ was applied. The background refractive index was set fixed to $n_{\mathrm{PBW,clad}} = 1.34$. Here, an elliptical taper cross-section has been chosen and X and Y diameters have been varied in order to find the optimum insertion loss.

Since slight misalignment of the PWB taper with respect to the SMF fiber-core might occur during fabrication, the misalignment tolerance of a linear circular PWB taper is simulated. Fig. 5.5 shows the overlap loss between the fundamental eigenmode of the PWB with a specific circular taper ending cross-section and the fundamental eigenmode of the standard single mode fiber when a certain x-y-offset between the fields is applied.

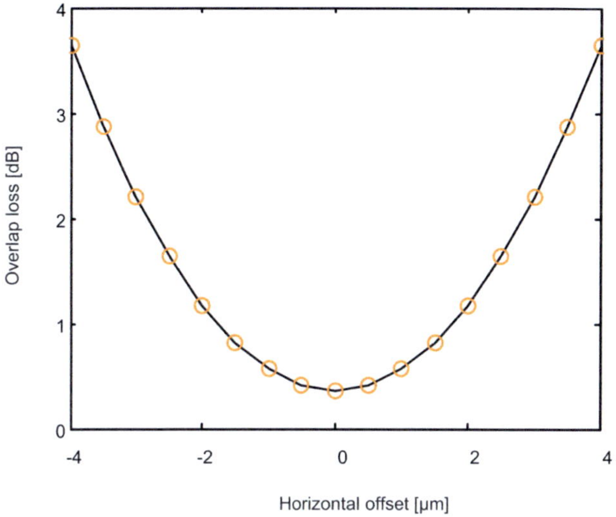

Fig. 5.5: Overlap mismatch loss for a horizontal offset between the fundamental eigenmode of the PWB with a circular ending cross-section of (diameter = 13.2 µm, which represents the optimum coupling for a 60 µm long taper) and the fundamental eigenmode of the standard single mode fiber. Even small misalignment results below 1 µm result in increasing losses. Therefore a careful alignment of the lithography system as well as a precise measurement of the fiber cores position is of vital importance.

As can be seen in the figure, overlap mismatch losses already start to become significant at offsets below 1 µm. Therefore, a careful adjustment of the lithography setup is indispensable.

5.2 Simulation of the Focal Field Distribution

From Chapter 1 it is known that coupling to vertical device facets is possible and laser power compensation can be used to ensure two-photon polymerization on the facet of metal-coated semiconductor devices. To identify the suitable power compensation, a similar simulation process as described in Section 4.2.1 is followed here. The vertical fiber facet is included into the simulation domain. In this setup, however, the size of the fiber is a major challenge. With a cladding diameter of d_{clad} = 125 µm and a core diameter of d_{core} = 8.3 µm the fiber is significantly larger than the semiconductor laser discussed in Chapter 1. Fig. 5.6 shows a schematic of the fiber inside and front view. For the simulation the single-mode fiber *Corning® SMF*-28 is modeled with a step index refractive index distribution of $n_{\text{core}} = 1.4589$ and $n_{\text{clad}} = 1.4537$ at a wavelength of $\lambda = 780$ nm. Since the actual values of *Corning® SMF*-28 at the desired wavelength are not reported by the manufacturer, the cladding is assumed to consist of pure SiO_2 [82] and the refractive index of the core is calculated from the known refractive index difference of 0.36%. Although these values might not represent the conditions in the SMF in an ideal way, they are considered as a sufficient starting point to understand the influence of the vertical fiber facet on the focal field distribution.

Due to limited computation resources the FDTD simulation of the whole fiber cross-section was not applicable. For this reason the fiber is scaled down and the perturbations introduced by the small fiber as a function of the distance between the geometrical focus and the vertical fiber facet are analyzed. All simulations were performed with a background refractive index of $n = 1.52$. The resolution was set fixed to 50 nm in all three dimensions and a numerical aperture of $NA = 1.3$ was used for the excitation field.

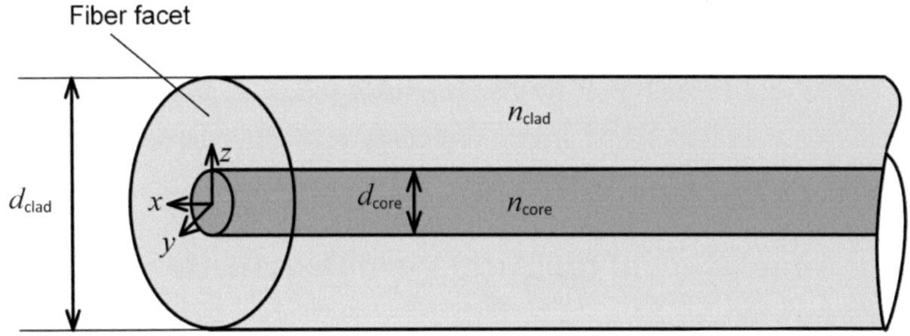

Fig. 5.6: Schematic of the standard single mode fiber: The fiber has a circular cross-section with a core diameter $d_{\text{core}} = 8.3$ µm and a cladding diameter of $d_{\text{clad}} = 125$ µm. The refractive indices used in simulations amount to $n_{\text{cladding}} = 1.4537$ and $n_{\text{core}} = (1+0.0036)\ n_{\text{cladding}} \approx 1.4589$ at $\lambda = 780$ nm. These values are calculated from the specified refractive index difference of 0.36 % between core and cladding and by assuming the fi ber cladding to consist of pure SiO_2.

The simulation was done for four different fiber radii r. In order to obtain comparable results the simulated squared, normalized electrical field $|E|^2$ is plotted versus the distance divided by the simulated radius. As can be seen in Fig. 5.7, starting from a fiber radius of 10.5 µm the simulation results for different fiber radii coincide which shows the convergence of the simulation. When the distance between vertical fiber facet and the undisturbed focus position is reduced to values smaller than the fiber radius, a decrease of the maximum field intensity $|E|^2$ can be observed. At a distance of $0.13r_{\text{fiber}}$ a minimum of the field intensity is reached at about 43% of the initial field intensity. When approaching the fiber facet further the maximum field intensity increases again.

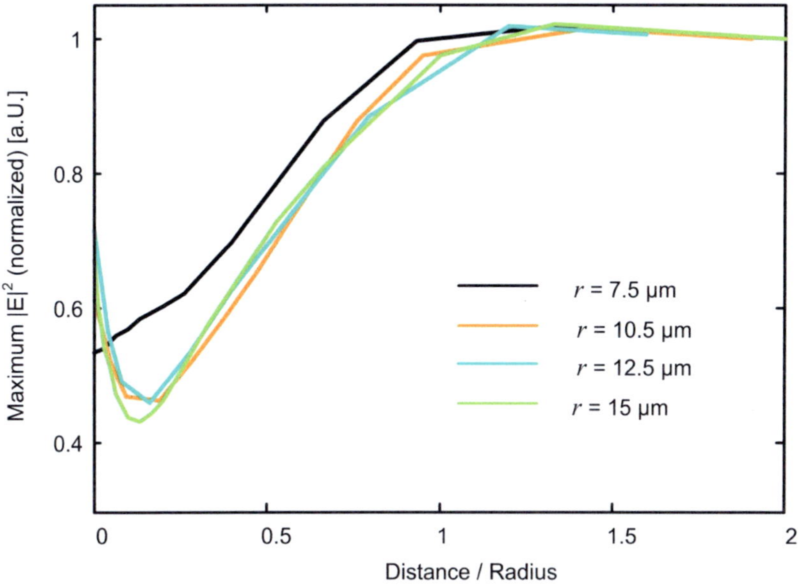

Fig. 5.7: Maximum normalized $|E|^2$ versus the distance divided by the simulated radius: when the fiber facet is approached to distances smaller than the fiber radius the maximum $|E|^2$ is reduced. The minimum $|E|^2$ is reached at a distance of approximately 0.13×r and, does increase when the distance is further reduced.

A visualization of the mechanism that leads to the increasing field intensity close to the fiber facet can be seen in Fig. 5.8. When the geometrical focus of the lithography system approaches the fiber facet the electrical field distribution develops additional maxima. The intensity in these maxima grows until a distance of about $0.13 r_{\text{fiber}}$. If the distance is further reduced, the intensity in the side lobes reduces and is transferred back to the main maximum.

To understand the origin of the side lobes observed in Fig. 5.8 in more detail, the propagation of the lithography beam through the simulation domain at certain time steps with the geometrical focus located in front of the fiber facet (r = 12.5 µm) at a distance d = 0.5 µm is investigated.

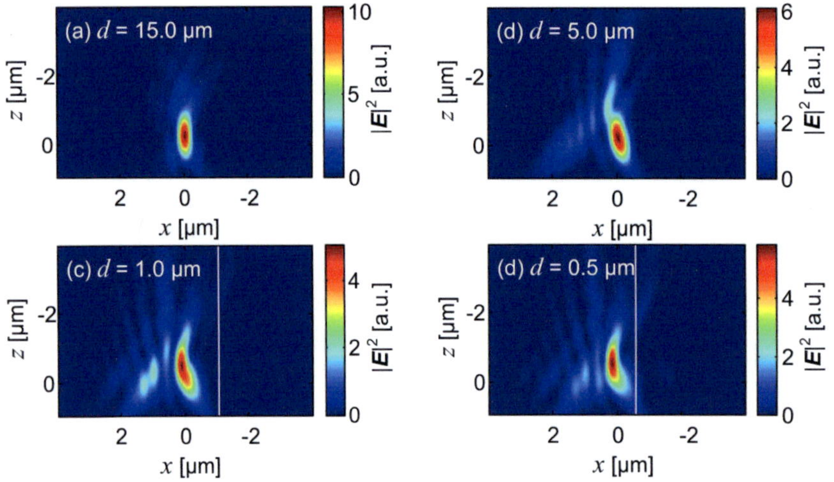

Fig. 5.8: xy-cuts of $|E|^2$ through the point of maximum $|E|^2$ at different distances (a) d = 15 µm, (b) d = 5.0 µm, (c) d = 1.0 µm, and (d) d = 0.5 µm between the geometrical focus and the vertical fiber facet. The fiber facet is indicated by the white line. The fiber radius is fixed at r = 15 µm: When the distance between the geometrical focus of the lithography system and the fiber facet is reduced the electrical field distribution develops additional maxima. The intensity in these side lobes grows until a certain distance. If the distance is further reduced the intensity in the side lobes reduces and is transferred back to the main maximum.

When entering the glass fiber, the lithography beam splits up into two parts that travel at different velocities due to the different refractive indices in the fiber and the background material. A similar effect would also be expected in a geometrical optics approach. The part of the focused laser beam that has entered the glass fiber leaves the front facet and leads to the formation of a second intensity maxima with almost equal field intensity (Fig. 5.9, e).

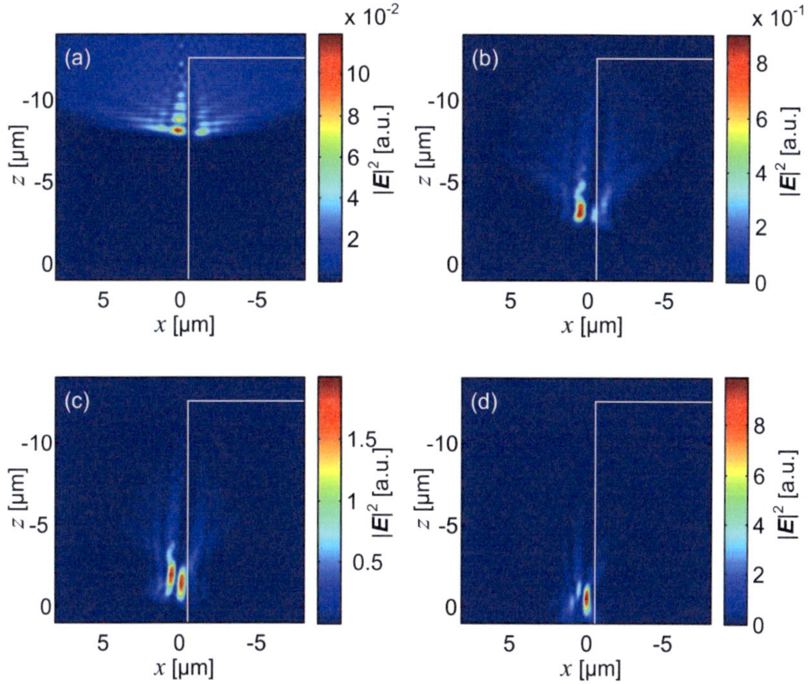

Fig. 5.9: xy-cut of $|E|^2$ in the position $y = 0$ after a certain amount of time steps (from (a) to (d): number of time steps = 400, 750, 900, 2000 = steady state criterion met). The edge of the fiber is indicated by a white solid line. When entering the glass fiber the lithography beam splits up into two parts that travel at different velocities due to the different refractive indices in the fiber and the background material. The part that has entered the glass fiber leaves the front facet and leads to the formation of a second intensity maximum with almost equal field intensity. In the following time steps the two maxima merge, interfere and parts of the second strong maximum form the side lobes found in Fig. 5.8.

In the following time steps, the two maxima merge and parts of the second maximum form the side lobes found in Fig. 5.8.

Due to the interference of the two different parts the point of maximum intensity no longer coincides with the undisturbed geometrical focus of the lithography system, which means that the actual focus of the system will shift and polymerization will not take place in the desired position. Fig. 5.10 shows the shift of the point of maximum $|E|^2$ with respect to the geometrical focus of the system as a function of distance between the geometrical focus and the fiber facet for a fiber radius $r = 62.5$ µm. When approaching the fiber, the point of maximum $|E|^2$ is shifted towards the lithography objective. At very small distances the ocal shift decreases again.

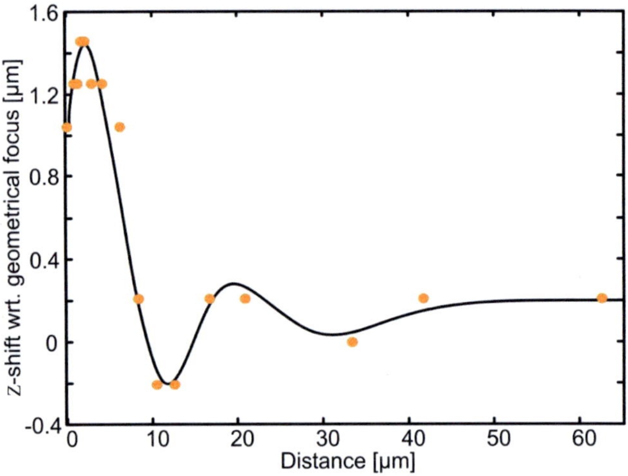

Fig. 5.10: Shift of the point of maximum $|E|^2$ with respect to the geometrical focus of the system as a function of the distance between the geometrical focus and the fiber facet calculated for the dimensions of a standard single mode fiber. At a distance larger than the fiber radius the geometrical focus of the lithography system and the point of maximum $|E|^2$ coincide. When approaching the fiber the point of maximum $|E|^2$ is shifted towards the lithography objective. At very small distances the focal shift decreases again. The maximum shift amounts to 1.46 µm.

For a fiber radius $r = 62.5$ µm, as it is found in a standard single mode fiber, this simulation indicate a maximum focal shift of approximately 1.5 µm towards the lithography objective in a distance of approximately 1.7 µm from the fiber facet.

5.3 Power Compensation

In order to compensate for the power loss close to the facet, a power compensation has been implemented that increases the laser power of the lithography system when approaching the fiber facet. Different compensation functions have been experimentally examined, where a cubic increase of the laser power as well as the laser power compensation according to the simulated power loss as shown in Fig. 5.7 have proven to lead to convincing results.

5.4 Taper Test

In the experiments, the single mode fiber Corning SMF28^{TM} with a core diameter of 8.3 µm, a cladding diameter of 125 µm and a refractive index difference of 0.36 % between the fiber core and cladding is used. The mode field diameter is specified to be 10.4 µm with tolerances ranging from 9.5 µm to 11.5 µm.

Before being able to connect PWB to the fiber facets, different preparation steps are necessary. After being stripped of its outermost polymer coating, the glass fiber is cleaved with a diamond cleaver to obtain a flat vertical device facet. The fiber is then cleaned using Ethanol and mounted on top of a glass carrier by adhesive bonding. An optical transparent UV curable adhesive (OP-4-20 632^{TM} from Dymax, [83]) is used. As a carrier, standard microscope slides with a thickness of approximately 1 mm are used. In the setup of the experiment it is essential to provide a transparent substrate since the used method to detect the fiber core position requires transmitted light.

After pre-positioning of the fiber, the entire sample comprising fiber and submount is rinsed in acetone and kept in a bath of isopropyl alcohol for minimum of five minutes. The sample is dried in a flow of N_2-gas.

In the next process step, the sample is clad with the resist material, and the port detection is performed inside the lithography system. While the port detection has been automated for SOI waveguides, a fiber-port-detection is not available at the present stage. For this reason, the fiber core detection is performed manually. It has been observed experimentally that the image lines indicating the position of the fiber core vanish when the fiber core is placed exactly in the focal plane of the camera system. In the case where it is slightly out of focus the core is, however, clearly visible when the fiber is observed under transmitted light illumination. This technique, also being used in fiber splicers [84], measures the axial position of the fiber core whereas the lateral position of the fiber facet and center of the core are determined visually. By measuring the lateral and axial position of the fiber core at the very facet and at a certain distance inside the fiber the fiber coupling position and direction are obtained.

Several taper structures are fabricated at the vertical cleaved facets of small fiber pieces in order to identify the most convenient lithography parameters for the fabrication of fiber-to-chip PWB. The fabricated taper structures have a starting circular cross-section with a diameter of $d_{\text{taper},2}$=13.6 µm (compare to Fig. 5.2). Except the taper depicted in Fig. 5.11 a the taper structures are fabricated in *IP-Dip*™ [39] in the line writing technique described in Section 2.4.2. A dip-in configuration is used as depicted in Fig. 1.4. The sample in Fig. 5.11. a is fabricated from SU-8 [38] since structure generation at a vertical fiber facet with a linear increase is not possible in *IP-Dip*™ [39], the resulting structures typically wash off during resist development. The laser power is been adapted in the area close to the vertical device facet according to different compensation curves.

As can be clearly seen from Fig. 5.11 suitable laser power compensation is possible. The most convincing results are found from the laser power compensation according to the simulated values shown in Fig. 5.7, but the use of quadratic or cubically increasing laser power towards the facet might also be useful in certain applications, e. g. to additionally adapt for decreasing laser power towards the edges of the writing field in the galvanometer based lithography system.

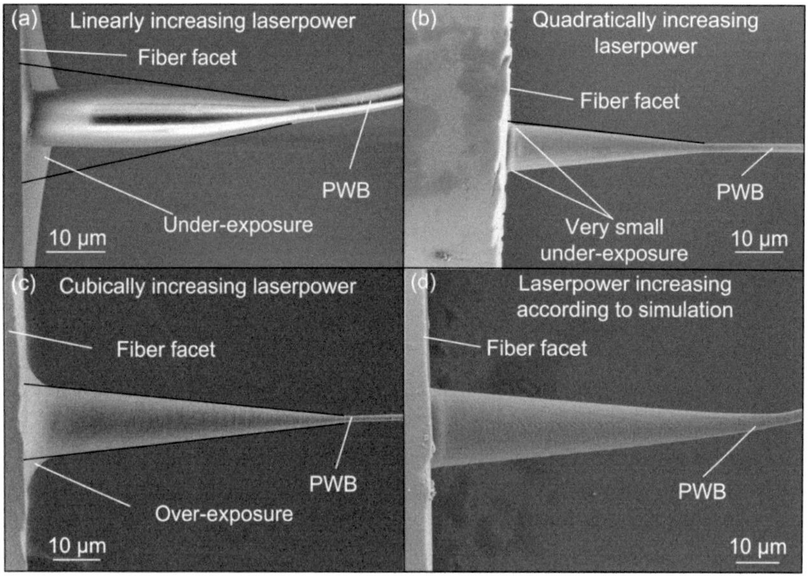

Fig. 5.11: SEM images of several taper test structures: the laser power close to the vertical fiber facet is increased to compensate for power loss introduced by shadowing of the lithography beam. (a) linear increase over a length of 60 µm. A clear underexposure is visible. (b) quadratic increase of the laser power over a length of 60 µm. The overall structural quality is very high only a very slight underexposure is still visible at a very small distance to the fiber facet. (c) cubic increase of the laser power over a length of 60 µm. Close to the fiber facet the structure is over-exposed and the sample geometry is bigger than designed. (d) Laser power compensation according to the simulated curve shown in Fig. 5.7. The structure quality is very high.

Fig. 5.12 shows the experimental demonstration of the focal shift predicted by simulations, see Fig. 5.10. Here a taper test structure without photonic wire bond is fabricated using the laser power compensation according to the simulation results to compensate for laser power loss. A SEM image, acquired under a very high angle is suitable to make the bottom side of the taper visible. The white arrow indicates the direction in which the optical axis of the lithography system was oriented during the lithography process. Shape deviations from the designed conical shape of the taper test structure are clearly visible due to the before mentioned shift of focus intensity towards the lithography objective. The effect is validated in several follow-up experiments. Although the focal shift is clearly visible it is not assumed to that strong optical impairments are caused by that small deviation, a compensation for focus shift has therefore not been implemented in the following experiments.

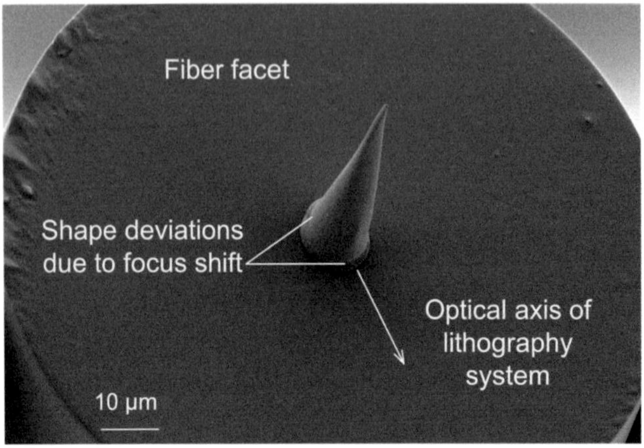

Fig. 5.12: SEM image of a taper test structure under an angle of 60°. Here a taper structure without photonic wire bond has been fabricated at the vertical fiber facet of a standard single mode fiber. The image has been acquired under a very large angle (60°) to make the bottom side of the taper visible. The white arrow indicates the direction in which the optical axis of the lithography system was oriented during the lithography process, with the arrow pointing away from the lithography objective. Shape deviations from the designed conical shape of the taper test structure are visible as they have been predicted by simulations, compare to Fig. 5.10. This focal shift has not been compensated for so far.

5.5 Fiber-to-Fiber-Coupling

A good way to identify the losses attributed to the interface between a PWB and a standard single mode fiber is the fabrication of a fiber-to-fiber PWB structure. This scheme is also used to demonstrate the applicability of fiber-coupling by photonic wire bonds in a planar configuration. Two SMF (cleaved under an angle of 0°) positioned side by side as well as face-to-face are connected *via* photonic wire bonding. The fabricated photonic wire bonds are shown in Fig. 5.13.

To ensure a low loss optical transmission, up-tapered sections as described in Section 5.1 and shown in Fig. 5.2 are used at the fiber interface. The length of the used up-tapered section amounts to $l_{PWB,taper} = 60\ \mu m$ with a taper end-diameter of $d_{taper,2} = 12\ \mu m$. The diameter of the un-tapered PWB waveguide section is designed as $d_{PWB} = 1.4\ \mu m$ in both experiments. The structures are fabricated from SU-8 [38] material in the line writing technique described in Section 2.4.2 and a cubic power increase towards the fiber facets is used during lithography to compensate for shadowing losses. In the case of the face-to-face device shown in Fig. 5.13. b the trajectory is designed longer than required to compensate for shrinkage. This kind of compensation, however, has proven error-prone since the resulting trajectory shows a strong bending that results in a low transmission.

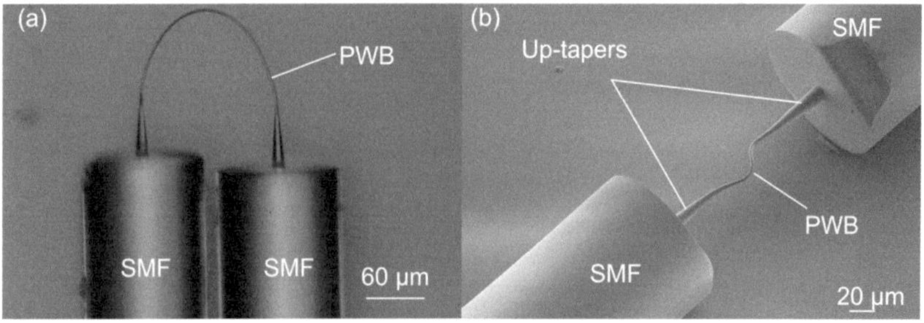

Fig. 5.13: Fiber-to-fiber connections by photonic wire bonding: The length of the utilized up-tapered section amounted to $l_{PWB,taper} = 60\ \mu m$ with a taper end-diameter of $d_{taper,2} = 12\ \mu m$. The diameter of the un-tapered PWB waveguide section has been designed as $d_{PWB} = 1.4\ \mu m$ in both experiments. The structures have been fabricated from SU-8 [38] material in the line writing technique. A cubic power increase towards the fiber facets was used during lithography to compensate for shadowing losses. (a) Optical image of a PWB connecting two fibers positioned side-by-side. The distance between the fiber cores amounts to 140 µm. The optical losses of this device amount to 2.5 dB at a wavelength of $\lambda = 1550$ nm. (b) SEM image of a PWB connecting two fibers positioned face-to-face. Here the trajectory was designed longer than required to compensate for shrinkage. This kind of compensation, however, has proven error-prone since the resulting trajectory shows a strong bending that results in a transmission of -6.8 dB at $\lambda = 1550$ nm.

To evaluate the optical losses of the fabricated structure, the setup depicted in Fig. 5.14 is used. Light from a tunable laser source is launched through a polarization controller into the device under test (DUT) through FC/APC connectors spliced to the feed-fibers. After the device under test light is detected in a powermeter as well as in an optical spectrum analyzer that is used to measure the wavelength dependent transmission of the DUT. The transmission through the setup without the DUT is used as a reference measurement. To mimic the lower index cladding material index oil with a refractive index of $n_{\mathrm{clad}} = 1.34$ is used in this experiment. The wavelength dependent transmission of the device displayed in Fig. 5.13. a is shown in Fig. 5.15. Measured insertion losses amount to 2.5 dB at a wavelength of $\lambda = 1550$ nm. Assuming the waveguide section of the PWB to be approximately loss less, this amounts to a taper loss of 1.25 dB in this experiment. This is in good agreement with the simulated losses that amount to 0.7 dB per taper in this case.

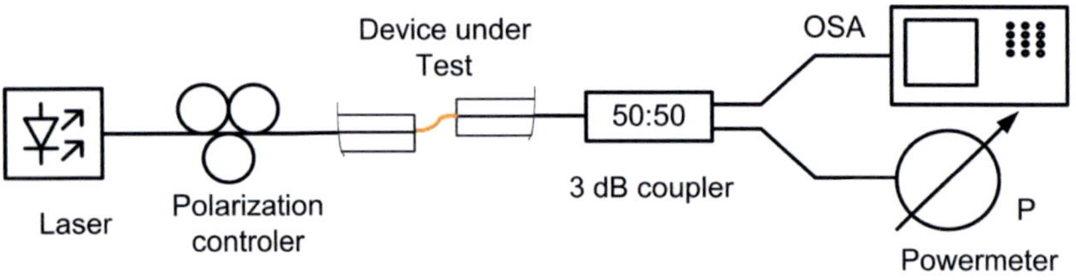

Fig. 5.14: General setup for broadband measurement of varies devices under test (DUTs). The laser emits light in a range from 1480 nm to 1550 nm, the polarization is controlled for optimum input. Half of the output power is measured by a powermeter, for real-time adjustments and the other half by an optical spectrum analyzer (OSA) for a fast broadband measurement.

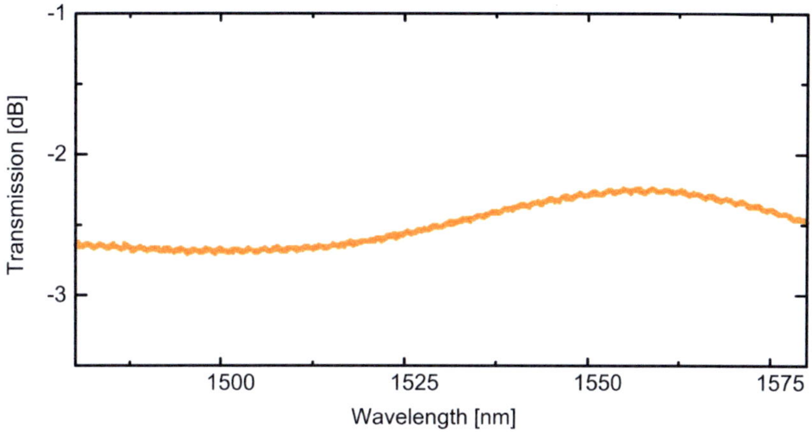

Fig. 5.15: Wavelength dependent transmission of the device displayed in Fig. 5.13. a. Measured insertion losses amount to 2.5 dB at a wavelength of $\lambda = 1550$ nm. Assuming the waveguide section of the PWB to be approximately loss-less, this amounts to a taper loss of 1.25 dB in this experiment which is in good agreement with the simulated losses that amount to 0.7 dB per taper in this case.

5.6 Coupling of Standard Single Mode Fibers to Silicon-on-Insulator Chips

In a fiber-to-chip coupling experiment the PWB interconnection between an SMF and a nanophotonic on-chip SOI waveguide are investigated.

As a demonstrator, a photonic wire bond connecting a nanophotonic SOI waveguide to a standard single mode fiber in a planar configuration is fabricated, as it is depicted in Fig. 4.3. The photonic wire bond bridges a distance of 22 µm in the y-direction; it spans 223 µm in the x-direction and overcomes a height offset of $\Delta h = 4.3$ µm. The PWB taper towards the fiber, as well as towards the SOI waveguide has a length of $l_{\text{up-taper}} = l_{\text{PWB,taper}} = 60$ µm. The taper end at the fiber interface has a diameter of $d_{\text{taper,2}} = 12$ µm with a circular geometry. The PWB taper at the SOI interface has a taper tip of $h_{\text{PWB,taper,1}} = 0.45$ nm in the vertical direction and $w_{\text{PWB,taper,1}} = 0.78$ µm in the horizontal direction. The PWB cross-section in the un-tapered region is designed to be $d_{\text{PWB}} = 2$ µm with a circular cross-section. Simulated coupling loss for the taper between the photonic wire bond waveguide section and the SMF amounts to approximately 0.65 dB. The structure is fabricated from *IP-Dip*™ [39] material ($n_{core} = 1.53$) in the helix writing strategy, according to the discussion in Section 2.4.2. A dip-in sample configuration as shown in Fig. 1.4 is used. A laser power compensation scheme is applied using a cubic increase of the laser power towards the fiber facet over a length of 60 μm. A SEM image of the fabricated device can be seen in Fig. 5.16.

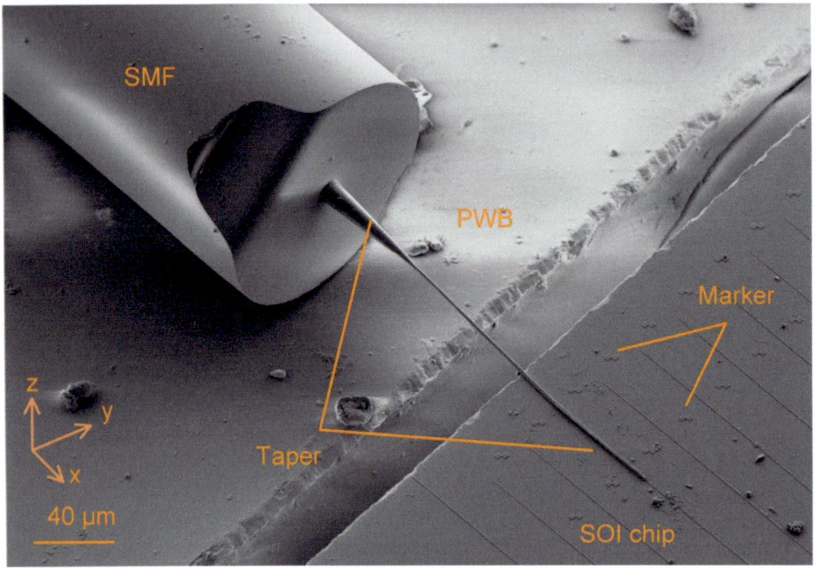

Fig. 5.16: SEM image of the fabricated fiber-to-chip PWB. The photonic wire bond bridges a distance of 22 μm in the x-direction; it spans 223 μm in the y-direction and overcomes a height offset of $\Delta = 4.3$ μm. The PWB taper towards the fiber as well as towards the SOI waveguide has a length of $l_{up\text{-}taper} = l_{PWB,taper} = 60$ μm, compare to Fig. 3.6 and Fig. 5.2. The taper end at the fiber interface has a diameter of $d_{taper,2} = 12$ μm with a circular geometry. The PWB taper at the SOI interface has at taper tip of $h_{PWB,taper,1} = 0.45$ nm in the vertical direction and $w_{PWB,taper,1} = 0.78$ μm in the horizontal direction. The PWB cross-section in the un-tapered region is designed to be $d_{PWB} = 2$ μm with a circular cross-section. The polymer waveguide core consists of *IP-Dip*™[39] ($n_{core} = 1.53$). Index matching liquid (Cargille, $n = 1.34$ at 1550 nm) is used as a cladding material. The structure has been fabricated in the helix writing technique described in Section 2.4.2.

To measure the optical insertion loss of the photonic wire bond, the transmission of the whole assembly is measured comprising the standard single mode fiber, the photonic wire bond, a straight SOI waveguide, a grating coupler as well as a second fiber for coupling out of the SOI waveguide again. The wavelength dependent transmission is shown in Fig. 5.17.

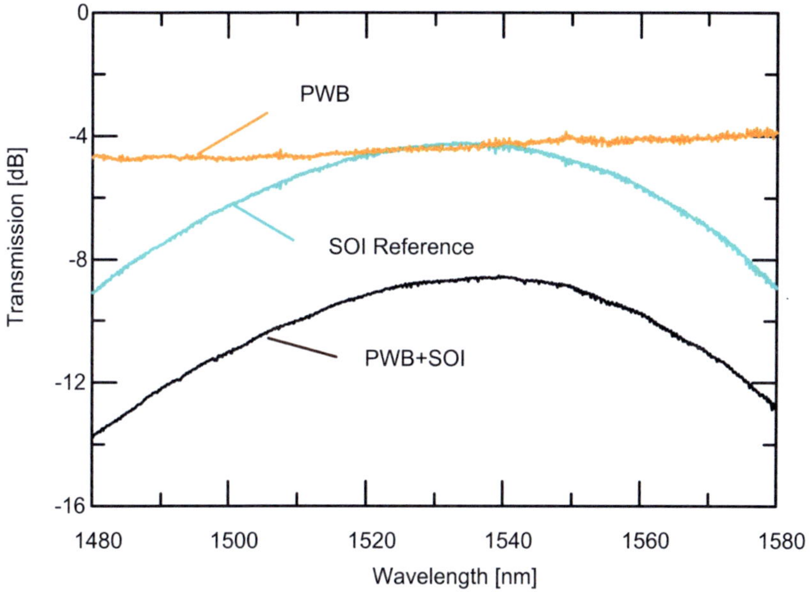

Fig. 5.17: Measured wavelength dependent transmission of the PWB device shown in Fig. 5.16. The solid red line represents the transmission of the whole assembly comprising the standard single mode fiber, the photonic wire bond, a straight SOI waveguide, a grating coupler as well as a second fiber for coupling out of the SOI waveguide. For comparison, an SOI reference configuration is measured comprising two fibers, two grating couplers and a straight SOI waveguide. Since the photonic wire bond device contains only one grating coupler, the reference configurations losses are divided by a factor of two (solid black line). The access loss of the PWB (green solid line) amounts to 4.2 dB at 1550 nm and shows a flatness of 1 dB over the whole measurement range from 1480nm to 1580nm

Index matching liquid (Cargille, $n = 1.34$ at 1550 nm) is used as a cladding material. For comparison, an SOI reference configuration is measured comprising two fibers, two grating couplers and a straight SOI waveguide. Since the photonic wire bond device contains only one grating coupler the reference configurations losses are divided by a factor of two to get a worst case estimate for the PWB access losses. The access loss of the PWB amounts to 4.2 dB at 1550 nm and shows a flatness of 1 dB over the whole measurement range from 1480 nm to 1580 nm, as can be seen in Fig. 5.17. This shows that the photonic wire bonds access loss can be lower than the coupling losses of standard grating couplers over a very broad wavelength range.

With this experiment, it is demonstrated that photonic wire bonds enable coupling between single-mode fibers and silicon-on-insulator waveguides. The measured optical losses of this planar fiber-to-chip interconnect, however, lack behind the expectations – although still being comparable to conventional fiber-to-chip coupling using grating couplers. This rather high insertion loss is attributed to a misalignment of the PWB with respect to the fiber core.

5.7 Multi-Core Fiber-to-Chip-Coupling

This section presents the fabrication and characterization of photonic wire bonds as an interconnection between a multi-core fiber and an SOI chip. This work has been published in Journal of Lightwave Technology [J1][1]. All fabrication details on the presented sample are summarized in Appendix A.6.

– Beginning of Paper [J1] –

Connecting Silicon Photonic Circuits to Multi-Core Fibers by Photonic Wire Bonding

Nicole Lindenmann, Stephan Dottermusch, Maria Laura Goedecke, Tobias Hoose, Muhammad Rodlin Billah, Temitope Onanuga, Andreas Hofmann, Wolfgang Freude, Senior Member, IEEE, Member OSA, Christian Koos, Member, IEEE, Member OSA

Abstract— Photonic wire bonding is demonstrated to enable highly efficient coupling between multi-core fibers and planar silicon photonic circuits. The technique relies on *in-situ* fabrication of three-dimensional interconnect waveguides between the fiber facet and tapered silicon-on-insulator waveguides. Photonic wire bonding can easily compensate inaccuracies of core placement in the fiber cross-section, does not require active alignment, and is well suited for automated fabrication. We report on the design, on fabrication, and on characterization of photonic wire bonds. In a proof-of-principle experiment, a four-core fiber is coupled to a silicon photonic chip, leading to measured coupling losses as small as 1.7 dB.

Introduction: Over the last years multi-core fibers (MCF) have proven to dramatically increase the transmission capacity of optical links by enabling

[1] The images in this sections have been reformatted for better readability, reference numbering and figure numbering have been changed for consistency.

space-division multiplexing (SDM) [85]. For exploiting the full potential of SDM in real-world transmission systems, MCF need to be connected to highly integrated photonic transmitter and receiver circuitry. The associated fiber-chip interfaces still represent a major challenge: Unlike the cores of the MCF, waveguides in photonic integrated circuits (PIC) are arranged in a common plane, and edge coupling is hence impossible. To overcome this problem, out-of-plane connection of a seven-core fiber to a silicon-on-insulator PIC has been demonstrated using an array of grating couplers [86]. This technique requires elaborate active alignment for adjusting the relative position of the MCF and the PIC in six degrees of freedom and is therefore not well suited for industrial mass production. Moreover, because of random deviations of the core positions within the cross section of the MCF, it is impossible to optimize the coupling efficiencies of all core-chip interfaces simultaneously. For this reason, a rather high insertion loss of 9.8 dB was measured [86]. In addition, out-of-plane coupling impedes the construction of flat packages, where fiber axis and chip surface are aligned. Furthermore, transmission bandwidth of grating coupler connections is intrinsically limited, making the concept unsuited for massively parallel wavelength division multiplexing schemes that use a large number of channels.

In this paper we show that photonic wire bonding [J2],[C6] provides a viable method for MCF-chip interfaces that overcome the limitations of grating couplers. Photonic wire bonding allows fabricating free-standing polymer waveguides with diameters down to 1 µm *in situ* by direct-write three-dimensional (3D) laser lithography based on two-photon polymerization. In previous experiments, we successfully demonstrated single-mode interconnects of silicon PIC on different chips, showed transmission at terabit/s data rates, and reported ways to reduce the transmission loss of photonic wire bonds (PWB) interconnecting silicon-on-insulator (SOI) waveguides (WG) to 1.6 dB. Here we expand on these results and show coupling of a four-core fiber to a silicon photonic chip with insertion losses down to 1.7 dB. The technique renders active alignment obsolete, can easily compensate tolerances of the core placement, and lends itself to automated mass production. Moreover, the fiber can be positioned in the same plane as the chip surface, thereby enabling greatly simplified and fully planar packages.

Designing the Photonic Wire Bond: Besides being single-moded at a wavelength of 1550 nm, photonic wire bonds have to fulfill a number of other requirements: Most important are small losses over a large bandwidth, a physically feasible trajectory without intersecting obstacles, and a mechanically stable shape that withstands the capillary forces during the removal of the unexposed part of the resist, from which the PWB is made of. Fig. 5.18 shows a schematic illustration of a photonic wire bond interconnecting one core of a MCF and an SOI chip. In this example, the photonic wire bond has tapered sections on the fiber endface as well as towards the SOI chip for adapting the mode fields. Between the tapered structures, the PWB has a round cross-section, and the PWB axis follows a 3D trajectory in space. The following section describes the optimization of the PWB trajectory, and the design of the tapers near the MCF as well as near the nanophotonic SOI waveguides.

Low-loss trajectory: For minimum optical loss, the trajectory of a PWB axis has to obey a number of constraints. To begin with, the endfaces of the PWB must overlap with the connecting waveguides on both sides. Next, the starting and ending orientation of the PWB axes have to coincide with the connecting waveguide axes. Further, the trajectory of the PWB axis is to be chosen such that intersections with obstacles like fibers, chip edges or other PWB are avoided. Finally, increased losses by a strong curvature of the trajectory and a large length should be avoided, and a suitable compromise has to be found. To fulfill the requirements listed above, a complicated optimization process has to be employed. Because the target (objective, cost) function is the optical transmission loss which has to be minimized, an appropriate loss model is needed for the PWB. While the theory of waveguide bends lying in a plane is well developed [44]–[46], [87], no theory is available for a waveguide with an axis which is bent in three dimensions. Work is in progress to find the parameters of a closed-form empirical function for the PWB losses. Meanwhile, with the help of full-wave finite-integration technique (FIT) calculations (CST Microwave Studio), we developed an expert system for designing the trajectories empirically, based on fifth-order polynomials.

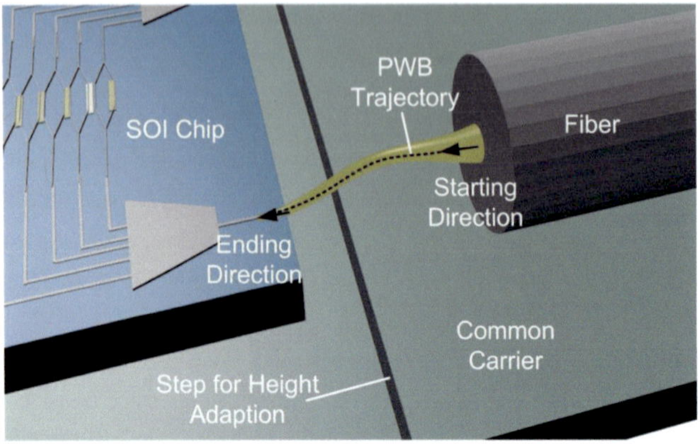

Fig. 5.18: Schematic of a photonic wire bond between fiber and silicon-on-insulator chip.The PWB trajectory is selected such that it adapts to the positions and the optical axes of the structures to be connected. Obstacles like chip or fiber edges must be avoided, and a good compromise is to be found between long interconnects and sharp waveguide bends.

Fiber interface: The fiber interface serves to convert the MCF mode field to the fundamental mode of the PWB. This interface is realized by a linear up-taper of the PWB waveguide; see Fig. 5.19. For designing the taper, we perform FIT simulations. We use a four-core fiber provided by Fibercore Inc. (SM-4C1500), featuring a step-index profile with a relative refractive index difference between core and cladding of $\Delta_{\mathrm{MCF}} = 0.54\ \%$ [88], a core diameter of $d_{\mathrm{MCF}} = 8\ \mu\mathrm{m}$, and a (Gaussian) mode-field diameter of 8.4 µm (1/e-diameter of the modal amplitude) as specified by the manufacturer. The PWB consists of a commercially available polymer with a refractive index of $n_{\mathrm{PWB,core}} = 1.53$ at a wavelength of 1550 nm (*IP-Dip*™ [39]). The diameter of the circular PWB cross-section is $d_{\mathrm{PWB}} = 2\ \mu\mathrm{m}$. The PWB is embedded in a low-index material that acts as an optical cladding as well as a mechanical support. We prefer *Cytop*™ [59] as a cladding material, having a refractive index of $n_{\mathrm{PWB,clad}} = 1.34$. The relative refractive index difference $\Delta = \left(n_1^2 - n_2^2\right) / \left(2n_1^2\right)$ between PWB core and cladding is $\Delta_{\mathrm{PWB}} = 13\ \%$, indicating a strong guidance.

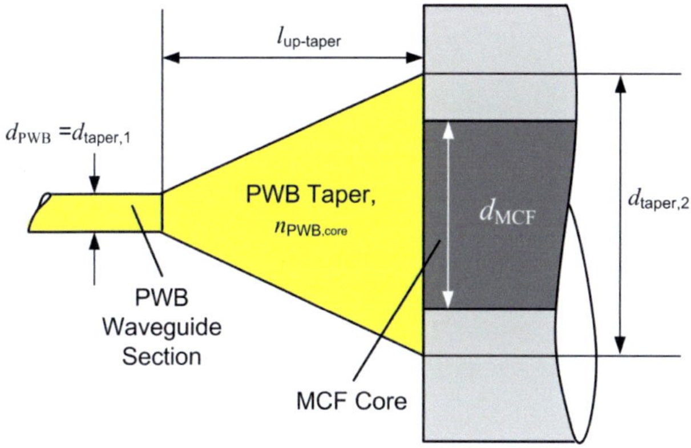

Fig. 5.19: Structure of the linear up-taper used at the interface between the MCF and the PWB.The taper starts with the circular cross-section of the PWB ($d_{\text{PWB}} = d_{\text{taper,1}} = 2$ µm) and ends with a circular cross section of larger diameter $d_{\text{taper,2}}$ at the fiber endface. This taper ending diameter is optimized for minimum insertion loss and turns out to be larger than the fiber core diameter $d_{\text{MCF}} = 8$ µm, because the relative refractive index difference between core and cladding is much larger for the PBW ($n_{\text{PWB,core}} = 1.53$, $n_{\text{PWB,clad}} = 1.34$, $\Delta_{\text{PWB}} = 13$ % at 1550 nm) than for the fiber ($\Delta_{\text{MCF}} = 0.54$ % at 1550 nm).

Parameter sweeps of taper length and taper diameter at the fiber endface result in a loss-optimized taper diameter as a function of the taper length, see Fig. 5.20. The shaded area in Fig. 5.21 shows the region where the minimum insertion loss (red line) increases by 0.1 dB. For a taper length of $l_{\text{up-taper}} = 60$ µm, a taper diameter range between 10.9 µm and 13.0 µm leads to a loss increase of less than 0.1 dB. This shows that the taper design is stable with respect to production tolerances and therefore well suited for fabrication. The simulated insertion loss of optimized tapers as a function of length is depicted in Fig. 5.20. The calculated loss between PWB and MCF amounts to 0.52 dB at a taper length of $l_{\text{up-taper}} = 60$ µm and taper diameter of 12 µm.

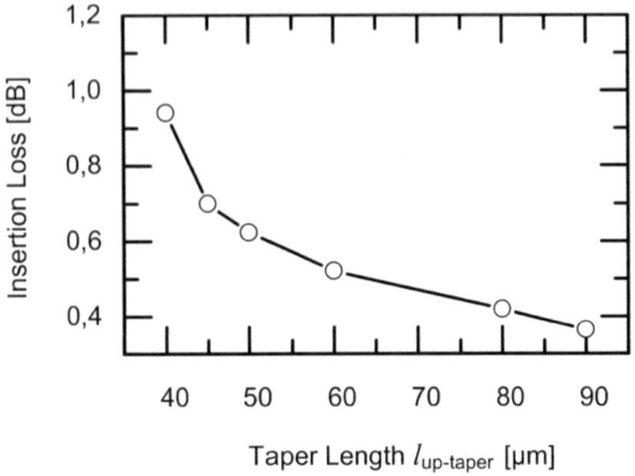

Fig. 5.20: Calculated insertion loss for optimized interfaces between the MCF and the PWB.A linear polymer taper with circular cross-section is used between the endface of the MCF (core diameter d_{MCF} = 8 μm, Δ_{MCF} = 0.54 % at 1550 nm) and the single-mode PWB section (core diameter d_{PWB} = 2 μm, core index $n_{\text{PWB,core}}$ = 1.53, cladding index $n_{\text{PWB,clad}}$ = 1.34 at 1550 nm). The diameter of the taper at the MCF interface is optimized for minimum insertion loss. The insertion loss is calculated between the fundamental eigenmode of the MCF and the fundamental eigenmode of the PWB waveguide section.

SOI chip interface: Next, we need to connect the PWB to a silicon-on-insulator strip waveguide that consists of a nanophotonic waveguide core on top of a thick buried oxide layer. An illustration of the interconnect is depicted in Fig. 5.23. For a low-loss transition between PWB and SOI WG we have to increase the mode field diameter of the SOI WG. This is done by laterally down-tapering the SOI WG towards the PWB ("inverse taper") [13], and by embedding the SOI WG into a rectangular polymer waveguide that is down-tapered towards the SOI WG. We refer to this structure as a 3D double-taper in the following. The 3D double-taper is covered by a cladding material. Fig. 5.23 shows the double-taper structure. The SOI WG tip has a width of w_{tip} = 80 nm. Measured from this tip, the investigated SOI WG taper lengths lie between l_{Si} = 40 μm and l_{Si} = 100 μm. The SOI WG taper connects to a straight SOI WG with a cross-section of w_{Si} = 500 nm (width) by h_{Si} = 220 nm (height). The PWB taper has a cross-section of $w_{\text{PWB,taper,1}}$ = 0.76 μm by $h_{\text{PWB,taper,1}}$ = 0.45 μm and ends with a cross-section of $w_{\text{PWB,taper,2}}$ = 2 μm by

$h_{PWB,taper,2}$ = 2 µm at the PWB waveguide side. The length of the polymer taper is fixed to $L_{PWB,taper}$ = 80 µm.

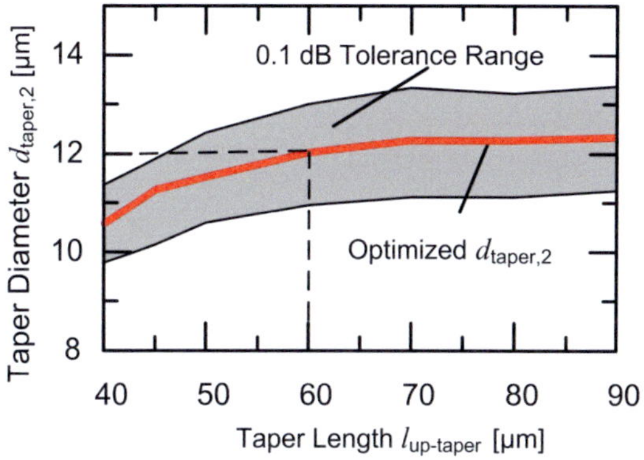

Fig. 5.21: Taper diameter $d_{taper,2}$ at the MCF interface vs. taper length $l_{up\text{-}taper}$ for a loss-optimized structure (red line). The linear polymer taper connects the single-mode PWB section (diameter d_{PWB} = 2 µm) with the endface of the MCF (SM-4C1500 core diameter d_{MCF} = 8 µm). At a taper length of 60 µm, the optimum taper diameter at the MCF interface amounts to $d_{taper,2}$ = 12 µm, and the minimum achievable transmission loss is 0.52 dB. In this case, the taper diameter $d_{taper,2}$ may vary between 10.9 µm and 13.0 µm for an increase of the insertion loss by at most 0.1 dB.

The rectangular PWB gradually changes its shape to a cylindrical PWB over a length of 20 µm. Polymer taper and PWB consist of the same material and are fabricated simultaneously.

The loss of the 3D double-taper is calculated between the fundamental eigenmode of the clad *rectangular* polymer taper WG and the fundamental eigenmode of the straight SOI WG. Again we employ a FIT solver, and we use the following additional data, all valid at a wavelength of 1550 nm: Refractive indices of silicon (n_{Si} = 3.48), silicon dioxide (n_{SiO2} = 1.44), PWB taper ($n_{PWB,core}$ = 1.53), and cladding ($n_{PWB,clad}$ = 1.34).

The silicon taper length varies between l_{Si} = 40 µm and l_{Si} = 100 µm. Except for the l_{Si} = 40 µm and l_{Si} = 100µm long SOI WG taper, the calculated losses lie between 0.75 dB and 0.9 dB in the C-Band (1530 nm – 1565 nm), see Fig.

5.22. Obviously, the 40 μm taper is too short for a low-loss transition. We attribute the ripple in the loss curves to multiple internal reflections inside the 3D double-taper.

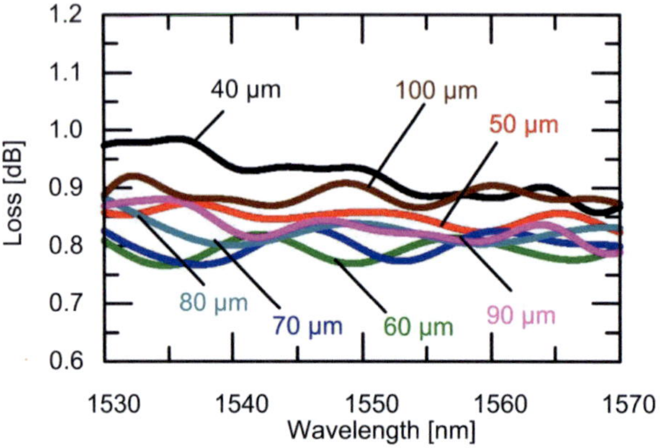

Fig. 5.22: Simulated transmission over wavelength for a 3D double-taper interface as shown in Fig. 5.23. The following values apply for the SOI waveguide: $w_{\mathrm{Si}} = 500$ nm, $h_{\mathrm{Si}} = 220$ nm, taper tip width $w_{\mathrm{tip}} = 80$ nm, refractive index $n_{\mathrm{Si}} = 3.48$ at 1550 nm. For the PWB taper: initial height $h_{\mathrm{PWB,taper,1}} = 450$ nm, initial width $w_{\mathrm{PWB,taper,1}} = 760$ nm, PWB waveguide section $w_{\mathrm{PWB,taper,2}} = 2$ μm × $h_{\mathrm{PWB,taper,2}} = 2$ μm, $n_{\mathrm{PWB,core}} = 1.53$, and $n_{\mathrm{PWB,clad}} = 1.34$ both at 1550 nm. The silicon taper length l_{Si} is varied in the simulation between $l_{\mathrm{Si}} = 40$ μm and $l_{\mathrm{Si}} = 100$ μm. Except of the $l_{\mathrm{Si}} = 40$ μm and $l_{\mathrm{Si}} = 100$ μm long SOI taper the simulation losses range between 0.75 dB and 0.9 dB.

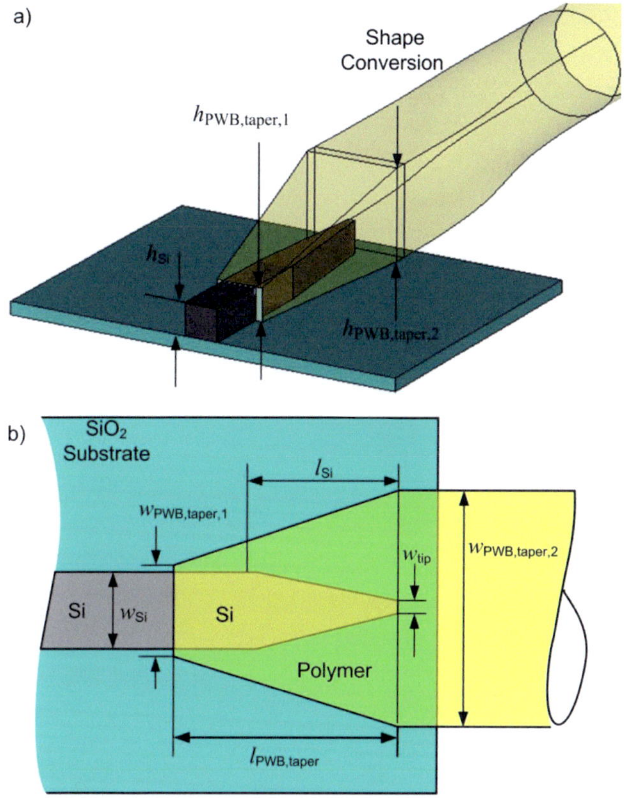

Fig. 5.23: 3D double-taper structure for modal field match between PWB and SOI wave-guide.(a) The SOI WG ends in a laterally down-tapered section ("inverse taper"), whereas the PWB features a section that is laterally and vertically down-tapered to adiabatically transform the SOI waveguide mode to that of the PWB. The PWB waveguide section gradually changes its cross-section from rectangular to circular. (b) The SOI WG tip has a width of $w_{tip} = 80$ nm. The SOI WG taper connects to a straight SOI WG with a cross-section of $w_{Si} = 500$ nm by $h_{Si} = 220$ nm. Measured from this tip, the investigated SOI WG taper lengths lie between $l_{Si} = 40$ µm and $l_{Si} = 100$ µm. The PWB taper has a rectangular cross-section of $w_{PWB,taper,1} = 0.76$ µm by $h_{PWB,taper,1} = 0.45$ µm and ends with a rectangular cross-section of $w_{PWB,taper,2} = 2$ µm by $h_{PWB,taper,2} = 2$ µm at the PWB waveguide section. The length of the PWB taper is fixed to $L_{PWB,taper} = 80$ µm. The lower-index cladding material is not displayed here for the sake of clarity.

Fabrication: For the realization of the photonic wire bond several fabrication steps are required. First, the objects to be interconnected (MCF and SOI WG in our example) are mounted on a common carrier using standard pick-and-place machinery with moderate precision. In this experiment we used a glass carrier with two different height levels for rough adjustment of the component heights, and we fixed the components with glue.

Second, the interconnect regions (MCF endface and SOI WG end) are embedded into a negative-tone photoresist, and the actual positions of the optical connection points (fiber core endfaces and taper tips of SOI WG) are detected by microscope image processing. With these data, the PWB shape (trajectory and waveguide cross-section along the trajectory) is calculated according to the criteria described previously.

Third, the calculated waveguide structures are defined by a direct-write 3D lithography technique. In our experiment, we use a commercial lithography system (*Photonic Professional* from *Nanoscribe GmbH,* [39]), which uses a frequency-doubled fiber laser emitting pulses at 780 nm wavelength with approximately 80 MHz repetition frequency and <100 fs pulse width. This laser light is focused into the resist through an immersion objective (100× magnification) with a large numerical aperture (NA = 1.3). The power at the input pupil of the objective amounts to approximately 4 mW. We use a high- resolution photo-resist (*IP-Dip*™), which is transparent at the laser wavelength, but enables two-photon polymerization in the high-intensity focal spot of the lithography system. Depending on the NA of the objective, on the laser power and on the sensitivity of the photo-resist, such a volume element (voxel) has a diameter (height) down to 100 nm (200 nm) [89]. In the lithography step, a refined two-stage writing strategy is used to define the PWB structures by moving the focal spot of the writing beam through the volume of the resist, see Fig. 5.24. The inner volume is exposed first by writing spiral lines, and an outer shell of contour lines is then added to smoothen the surface.

As a last fabrication step, the unexposed resist material is removed in a developer bath, and the resulting PWB is finally embedded in a low-index cladding material.

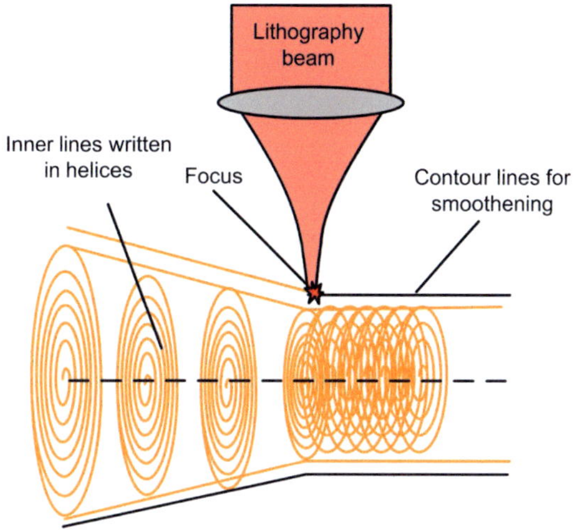

Fig. 5.24: Two-step writing strategy for PWB fabrication.The inner volume is exposed first by writing spiral lines. In a second step, an outer shell is added by writing lines parallel to the contour of the PWB for smoothening the surface.

In our experiments we immerse the photonic wire bond into an index-matching liquid with a refractive index of 1.34, but a low-index cladding material like *Cytop*™ would be preferred for stabilization.

For practical application of photonic wire bonding in manufacturing, writing speed is a crucial parameter. We are currently using a system, in which the lateral movement of the lithography beam is steered with galvanometer-driven scanning mirrors. For this system, achievable writing times amount to less than 5 min for a fiber-chip PWB. This is much faster than our conventional system, in which the focus of the laser beam was held in a fixed position and the high-inertia sample holder had to be moved by a piezo-driven translation stage, leading to writing speeds of approximately 75 µm/s and writing times of more than 1h per fiber-chip PWB. The structures presented below have still been written using the piezo system. For scanner-based lithography systems, we expect that further optimization will allow writing times of a few seconds per fiber-chip PWB in the future, making the technique suited for high-volume mass production.

Multi-core Fiber-chip Connection: Fig. 5.25 shows the resulting PWB interconnection between a four-core fiber and a silicon photonic chip. The interface consists of four PWB that connect the individual cores of the MCF to different on-chip waveguides.

At the interface towards the SOI WG, polymer tapers of length $L_{\mathrm{PWB,taper}} = 80$ µm are used. Their initial cross-sections are rectangular with $w_{\mathrm{PWB,taper,1}} = 0.76$ µm (width) and $h_{\mathrm{PWB,taper,1}} = 0.45$ (height), and they end with a quadratic cross-section of equal width and height, $w_{\mathrm{PWB,taper,2}} = h_{\mathrm{PWB,taper,2}} = 2$ µm at the PWB waveguide section. The embedded Si tapers have a length of $l_{\mathrm{Si}} = 60$ µm and a taper tip width of $\mathrm{w}_{\mathrm{tip}} = 80$ nm as described in Section I C. The PWB tapers towards the fiber endface have a starting diameter of $d_{\mathrm{taper,2}} = 12$ µm and a taper length of $l_{\mathrm{up\text{-}taper}} = 60$ µm The circular PWB waveguide sections have diameters of $d_{\mathrm{PWB}} = 2$ µm as described in Section I B.

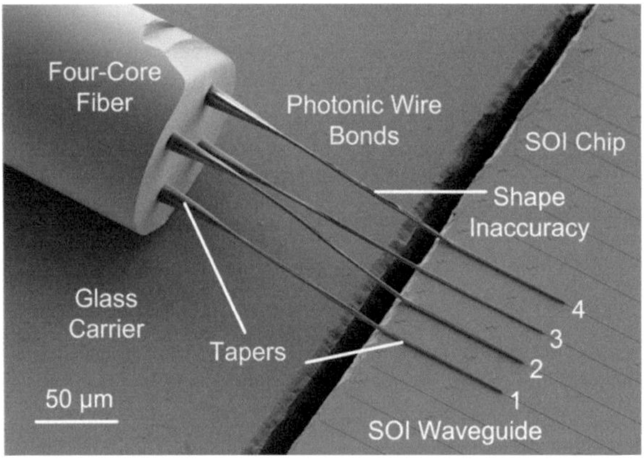

Fig. 5.25: Fabricated sample.Photonic wire bonds (PWB) connect the individual cores of a four-core fiber to different on-chip SOI waveguides. The PWB are up-tapered both on the MCF and on the SOI WG side to match the mode diameter to that of the fiber core and of the SOI WG, respectively. The PWB consist of a negative-tone photo-resist. At PWB 4, shape imperfections can be seen.

For measuring the insertion losses of the MCF-chip interconnects, we use the setup depicted in Fig. 5.26. Light is coupled to the individual SOI waveguides *via* grating couplers, and then passed on to the MCF cores *via* the PWB interface. The cleaved output facet at the far end of the MCF is scanned by a lensed fiber to measure the power distribution in the cross-section as depicted in Fig. 5.27. The best performance is found if light is launched into waveguide number 2. In this case we measure a total loss of 9.7 dB for an optimum alignment of the lensed SMF. Taking into account insertion losses of 4.8 dB at the grating-coupler interface and another 3.2 dB of coupling loss between the MCF core and the lensed fiber, we find a net insertion loss of 1.7 dB for the PWB interface. The losses of the grating-coupler interface were determined from a reference measurement, revealing a loss of 9.6 dB when coupling light through a simple SOI WG with two grating couplers. Similarly, the insertion loss between the MCF core and the lensed fiber has been obtained from a reference transmission experiment comprising a pair of lensed fibers and a piece of MCF. Applying the same technique to the remaining bonds 1, 3, and 4, we find insertion losses of 3.5 dB, 2.5 dB, and 6.8 dB, respectively.

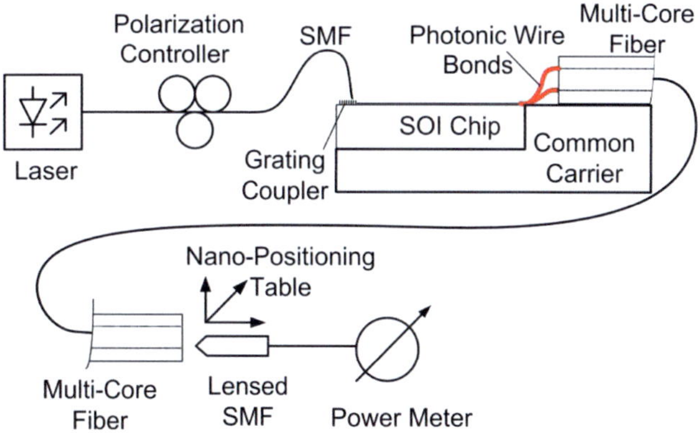

Fig. 5.26: Measurement setup.Laser light is coupled to the individual SOI waveguides by grating couplers, and then passed on to the MCF *via* the PWB connection. The cleaved output facet of the MCF is scanned by a lensed fiber to measure the power distributions.

All measurements were performed under usual laboratory conditions without special temperature control. The performance of photonic wire bonds under more challenging environmental conditions such as high temperatures and higher humidity is subject to ongoing research. In first tests we found that PWB maintain performance after being baked at 85°C at normal atmosphere for one hour, and multiple rinses in water also did not worsen the transmission of PWB. We therefore expect the structures to exhibit stability properties that are well suited for a wide range of practical applications.

Simulated losses of the fabricated structure amount to 1.3 dB, where 0.78 dB are due to the transition between the PWB and the SOI WG, and the remaining 0.52 dB are caused by the interface to the MCF. This compares well to the smallest measured insertion loss of 1.7 dB – the additional losses are attributed to scattering losses in the PWB waveguide section itself, caused by fabrication imperfections which cannot be included in the simulations, see Fig. 5.25. Moreover, shrinkage of the resist after exposure led to tension within the PWB, see Fig. 5.25. We believe that transmission and uniformity of the MCF-chip interfaces can be significantly improved in the future by optimizing taper structures and fabrication processes, leading to insertion losses of less than 1 dB or a fiber-chip interface.

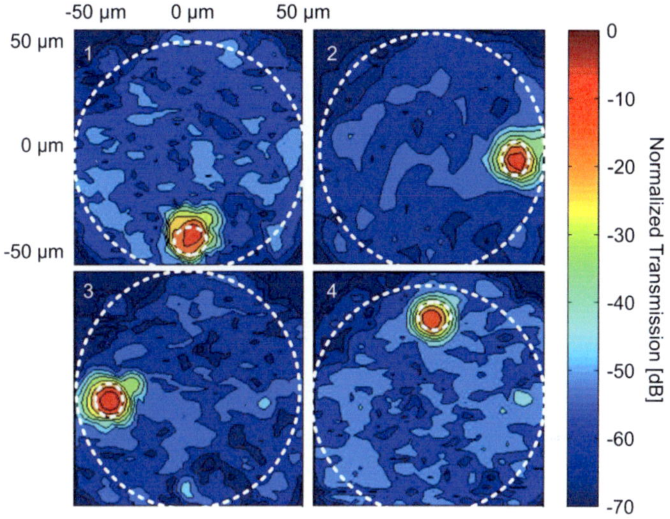

Fig. 5.27: Spatially resolved normalized power transmissionat the cleaved output facet of the multi-core fiber. The transmission is normalized to the maximum power level in each core, measured when launching light to the individual on-chip waveguides 1…4. These data correspond to insertion losses of 3.5 dB (PWB 1), 1.7 dB (PWB 2), 2.5 dB (PWB 3) and 6.8 dB (PWB 4). The rather high insertion loss of PWB 4 is attributed to fabrication imperfections, see Fig. 5.25 The net insertion loss of the MCF-chip interface is obtained by taking into account the losses of the grating-coupler interface (4.8 dB) and the coupling losses between the MCF-core and lensed fiber (3.2 dB).

Summary: We show that photonic wire bonding enables low-loss interfaces between multi-core fibers and single-mode SOI waveguides. The technique does not require active alignment and allows connecting a wide range of fiber types and core configurations. Insertion losses as down to 1.7 dB were measured, with much potential for further improvement.

– End of Paper [J1] –

6 Summary and Future Work

Photonic wire bonding has proven to be a viable candidate for future multi-chip optical assembling. The concept can combine the strength of several material platforms and can therefore be an alternative to monolithic integration of optical functionality on a common photonic integrated circuit.

In this thesis the process of photonic wire bonding has been developed. Proof-of-principle devices have been fabricated and their performance has been evaluated in numerous experiments. Methods have been developed to adapt the photonic wire bonds to different photonic integration platforms such as integrated SOI devices and semiconductor lasers. First steps towards automation have been taken, such as automated port detection and trajectory planning. By means of numerical simulations and compensation of the lithography beam coupling of photonic wire bonds to face-emitting devices such as face-emitting lasers and glass fibers has been enabled. In this chapter the key results are summarized and suggestions for future research are given.

Photonic wire bonding technology development:

In the framework of this thesis, photonic wire bonding as a novel photonic integration platform has been developed. The basis of this technique is multi-photon induced polymerization that happens in the focus of a strong, pulsed fs-laser that is focused under a high numerical aperture. By the 3D resolution of this method it becomes possible to generate 3D freeform polymer waveguides that connect to different photonic components on different substrates. In the framework of this thesis the technology of photonic wire bonding was developed, comprising all the necessary fabrication steps. First steps towards automation have been taken: Marker-based pattern-recognition was developed for automated port detection and a 3D routing algorithm was implemented for the calculation of the PWB trajectory. Algorithms have been developed creating a 3D bond structure around the said trajectory, incorporating dedicated taper structures for optimized coupling to and from different photonic components. Different writing strategies have been developed and the concept has been validated by prototype fabrication. It has been shown that photonic wire bonding provides a scalable way to interconnect different photonic material platforms.

With this newly developed technology it becomes possible to generate photonic multi-chip systems from known-good devices without the limitations given by monolithic integration.

Interconnections between SOI nanophotonic waveguides:

Off-chip connections between silicon photonic components and other devices mostly rely on out-of-plane coupling using grating couplers. Although these couplers provide optical access anywhere on the chip surface they introduce a number of drawbacks amongst which the high coupling angle of almost 90° and the limited optical bandwidth are the most severe ones. By the use of 3D double taper structures as integral part of the photonic wire bonds it was possible to demonstrate coupling between SOI waveguides with losses as low as 1.6 dB. Optical transmission over a bandwidth of 310 nm (from 1270 nm – 1580 nm) has been demonstrated with average losses of 2.5 dB with a standard deviation of 1.1 dB. Record high transmission bandwidth have been experimentally shown, proofing flawless optical data transmission up to 5.25 Tbit/s in a highly parallel wavelength-division multiplexing.

Calculation of optical fields in the focal region and coupling to vertical device facets:

To overcome the lack of active components for laser light emission in silicon photonics, coupling to standard laser devices is one of the key requirements for photonic wire bonding. Since most of those active components are manufactured as face-emitting components coupling to these devices inherently requires the fabrication of photonic wire bonds at vertical device facets. As parts of the lithography laser beam are disturbed by the presence of this vertical device facet, a compensation of the laser beam is required to enable structure generation. To this end a theoretical model was derived and a simulation environment was developed to numerically simulate the influences of the said devices facets on the focus quality. A writing strategy was developed that pre-compensates the focus power losses. With this technique proof-of-principle device fabrication of laser-to-chip coupling has been demonstrated. However, shrinkage of the used resist material is a major issue here that has to be addressed in the future.

Fiber-to-Fiber and Fiber-to-Chip Interconnects:

While lots of photonic functionality can be realized on-chip the interconnection between photonic systems and the outside world vitally requires the connection to an optical fiber. The interconnection between SMF and SOI nanophotonic waveguides in a planar configuration have been shown using photonic wire bonding. Alike in the case of SOI-to-SOI coupling, active alignment is obsolete here and the achievable optical bandwidth greatly exceeds the one provided by grating couplers.

Multi-Core Fiber Connection to On-Chip Waveguides:

First MCF-to-chip connections have been shown based on grating couplers [86] in an out-of-plane configuration using active alignment. This work, however, has overcome the inflexible setup of grating couplers that cannot be adapted to tolerances in core placement inside the multi-core fiber cross-section. Here, the fabrication of photonic wire bond interconnection between nanophotonic SOI waveguides and the individual cores of a four-core fiber has been demonstrated with excess losses as low as 1.7 dB. The single photonic wire bonds have been directly adapted to the actual positions of the fiber cores and can thus compensate for core placement tolerances.

Photonic wire bonding in the future:

To improve the photonic wire bond process further a number of potential future steps are proposed:

In many applications it is undesired to immerse the entire assembly of photonic components into a resist material. Possible reasons for this are a chemical incompatibility between the components and the lithography resist as well as the danger of sample pollution. To overcome the need to immerse the entire assembly the use of a **dispensing unit** that locally deposits the resist material only between the interconnect positions of the single components might be useful. Such a dispensing unit would also significantly reduce the amount of chemicals required for sample development.

For future automation the use of a **confocal scan** unit is recommended to obtain a direct 3D landscape of the pre-positioned photonic components. With this technique it would be possible to further improve accuracies since a potential misalignment of the camera system with respect to the lithography system does not any more influence the measured sample coordinates.

A closed-form model of the photonic wire bonds **trajectory related losses** is not available at the present stage. Based on theoretical findings on polarization dependent bending loss, torsion loss as well as transition losses it is suggested to extend the presented optimization algorithm by these loss contributions.

As it has been stated in Section 2.4 the lithography process of photonic wire bonding requires single writing lines to be connected to the already-exposed material or chip surfaces. One way to ensure such a secure **anchoring in line writing** technique would be to sort the individual writing lines according to their position in the PWB trajectory instead of their ascending order along the optical axis.

Expanding on the simulation environment developed in the framework of this thesis an algorithm might be developed to not only **pre-compensate** for laser power losses introduces by vertical device facets but also on the focal shape and position in sample configurations where parts of the lithography beam are disturbed by sample edges. Simulation results might be used to find the transfer function of the used sample configuration. With this transfer function it should then be possible to modify the input parameter, i. e. the optical field before the lithography objective such that the systems focus lies in the desired position and encounters the desired shape despite the influence of sample edges or partial laser light reflection and deflection. In this context, the use of a **spatial light modulator** inserted into the lithography system might be useful.

Apart from this, a spatial light modulator could also be used in sample generation where edges and surfaces do not worsen the focus quality. In those configurations, the input field could be modified to broaden the lithography focus in areas where high resolution is not required. By this method the required line density could be reduced and the lithography process could be accelerated.

As shrinkage of the resist material during sample development has proven to be a severe stumbling block, the development of a numerical model is required that can be used for the **simulation of shrinkage** and thus for the development of shrinkage compensation. This task, however, is considered to be very challenging since a closed-form dose model of the resist as a necessary precondition is not available at the present stage.

Appendix: Supplementary Material

A.1 Stl-File Export

The stl-format is used for data exchange between the PWB generation implemented in Matlab and the slicing tools provided by Nanoscribe (Nanoslicer, DeScribe), it is defined by surface triangles and the corresponding surface normal of the triangle in the following syntax:

```
facet normal -0.000000e+000 -1.000000e+000 -0.000000e+000
   outer loop
     vertex  1.558263e+002 1.500030e+002 1.054894e+002
     vertex  1.556642e+002 1.500030e+002 1.056565e+002
     vertex  1.559835e+002 1.500030e+002 1.053177e+002
   endloop
endfacet
```

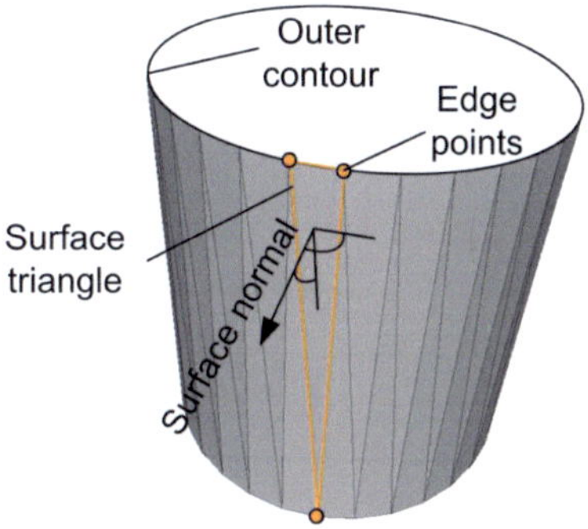

Fig. A.1: Export of the PWB in stl-format. The node points of the outermost PWB layers are sorted into triangles. For each triangle the normal is calculated from the cross product of the vectors between the tree edge points.

To generate this output file, the node points of the outermost PWB layers are arranged into triangles, as can be seen in Fig. A.1. For each triangle the normal is calculated from the cross product of the vectors between the tree edge points.

A.2 Fabrication Details [C21]

Table A.1: Fabrication Details for the PWB sample shown in Chapter 3.1 and published in [C21].

Substrate: SOI waveguide	
Width SOI WG	$w_{si} = 500$ nm
Height SOI WG	$h_{si} = 220$ nm
Coupling structures on chip:	
Grating coupler	
Grating patch:	12 µm × 15 µm
Material and sample perperation:	
PWB material:	MicroChem SU-8 2075
Refractive index core:	n_{co}=1.57 @ 1550 nm
Cleaning:	Bath in acetone Bath in isopropyl alcohol, Plasma-cleaning step
Applied:	Spin Coating
Prebake:	65°C 3 Minutes, 95°C 10 Minutes
Setup:	
Sample setup:	Sandwich, compare Fig. 1.3
Lithography setup:	Piezo-based, see Section 1.3.3
Writing strategy: Trajectory writing, see Section 2.4.1	
Distance between lines:	$\Delta x = \Delta z = 0.5$ µm
Velocity:	50 µm/s
PWB parameters:	
Trajectory:	semi-circle with radius of 53 µm
Cross-section:	circular
Diameter:	$d_{PWB} = 9$ µm

A.3 Fabrication Details [C20]

Table A.2: Fabrication details on the chip-to-chip PWB shown in Section 3.2.2 and published in [C20].

Substrate: SOI waveguides on two separate Chips	
Width SOI WG:	w_{Si} = 300 nm
Height SOI WG:	h_{Si} = 220 nm
Coupling structures on chip: Waveguide taper	
Taper length:	20 μm
Taper tip width	w_{tip} = 100 nm.
Material and sample preparation:	
Chip mounting:	Silver filled epoxy, EPO-TEK® H20E Epoxy Technology Inc. [80]
Adhesion promoter:	Ti Prime [90]
PWB material:	MicroChem SU-8 2075
Refractive index core:	n_{co}=1.57 at 1550 nm
Cladding:	Index Oil
Refractive index cladding:	n_{clad} = 1.34 at 1550 nm
Cleaning:	Bath in acetone Bath in isopropyl alcohol Dried in nitrogen flow, Plasma cleaning step
Applied:	Spin coating
Prebake:	65°C 5 Minutes, 95°C 45 Minutes

Setup:	
Sample setup:	Sandwich, see Fig. 1.3
Lithography Setup:	Piezo-based, see Section 1.3.3
On-chip positioning: Manual measurement of interface, no markers	
Writing strategy: Trajectory writing, see Section 2.4.1.	
Distance between lines:	Δx= 0.4 µm, Δz= 0.7 µm
Velocity:	50 µm/s
PWB parameters:	
Trajectory:	Forth order polynomial
Apex:	30 µm
Height difference:	12 µm
Cross-section:	Rectangular
PWB width:	$w_{\mathrm{PBW,taper,2}} = 2$ µm
PWB height:	$h_{\mathrm{PWB,taper,2}} = 1.6$ µm
PWB taper tip width	$w_{\mathrm{PWB,taper,1}} = 0.76$ µm
PWB taper tip height	$h_{\mathrm{PWB,taper,1}} = 0.45$ µm

A.4 Fabrication Details [C16]

Table A.3: Fabrication details of the low loss PWB shown in Section 3.2.3 and published in [C16].

Substrate: SOI waveguides on common chip	
Width SOI WG:	w_{Si} = 500 nm
Height SOI WG:	h_{Si} = 220 nm
Coupling structures on chip: Waveguide taper	
Taper length:	$l_{\mathrm{PWB,taper}}$ = 100 µm
Taper tip width	w_{tip} = 800 nm.
Material and sample preparation:	
Adhesion promoter:	Ti Prime [90]
PWB material:	MicroChem SU-8 2075 [38]
Refractive index core:	n_{co}=1.57 at 1550 nm
Cladding:	Index oil
Refractive index cladding:	n_{clad} = 1.3 at 1550 nm
Cleaning:	Bath in acetone Bath in isopropyl alcohol Dried in nitrogen flow Plasma cleaning step
Applied:	Spin Coating
Prebake:	65°C 5 Minutes, 95°C 45 Minutes
Setup:	
Sample setup:	Sandwich, See Fig. 1.3
Lithography setup:	Piezo-based, see Section 1.3.3.

Port detection algorithm as described in Section 2.1.	
Writing strategy:	Trajectory writing, see Section 2.4.1.
Distance between lines:	Δx= 0.4 µm, Δz= 0.7 µm
Velocity:	50 µm
PWB parameters:	
Trajectory:	Fifth order polynomial
Cross-section:	rectangular
PWB width:	$w_{\mathrm{PBW,taper,2}} = 2$ µm
PWB height:	$h_{\mathrm{PWB,taper,2}} = 1.6$ µm
PWB taper tip width	$w_{\mathrm{PWB,taper,1}} = 0.76$ µm
PWB taper tip height	$h_{\mathrm{PWB,taper,1}} = 0.45$ µm

A.5 Simulation of the Interface between Silicon-on-Insulator Waveguides and Photonic Wire Bonds

For all simulations a commercially available fully vectorial time-domain solver (CST Microwave Studio, [61]) is used. The simulation mesh was set to 25 lines per wavelength. A double taper structure already introduced in Chapter 1 is used as a mode converter between the ground mode of the SOI and PWB waveguides. All necessary dimensions are given in Fig. A.2. The following sections will discuss the influence of the single dimensions on the overall taper loss and performance.

A.5.1 Single Mode Dimensions

To maintain single mode operation of the PWB interconnect section the port modes of a rectangular PWB waveguide on top of a buried oxide substrate layer, covered by a lower index cladding material are simulated. The dimensions of the waveguide are varied and the port modes for each variation are calculated. If the waveguide exceeds its single-mode dimensions the first higher order mode becomes guided and its effective refractive index becomes higher than the refractive index of the surrounding cladding material. The simulation is performed for two different resist materials that have been used in the framework of this study, first for SU-8 with a refractive index of 1.57 at wavelength of 1550 nm [50] as well as for *IP-Dip*™[39] with a lower refractive index of 1.53.

For SU-8 single mode conditions are found to be fulfilled in a range indicated by Fig. A.3a.

For *IP-Dip*™[39] the single mode conditions are more relaxed since the index contrast between core and cladding of the waveguide is smaller. Fig. A.3 (b) depicts the area of single mode operation for *IP-Dip*™[39] as a waveguide material and a rectangular waveguide cross-section.

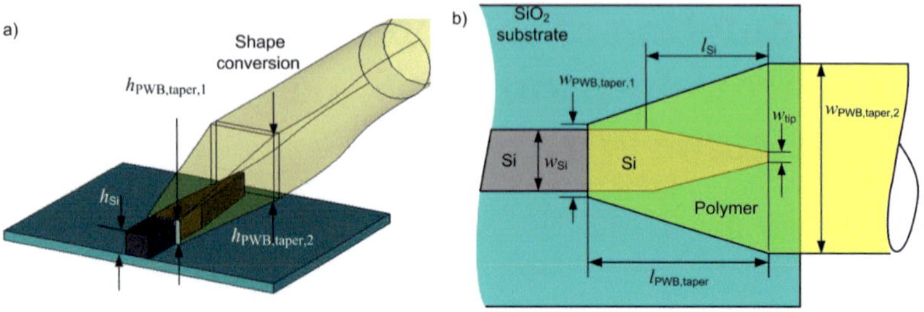

Fig. A.2: 3D Double-taper structure for modal field match between PWB and SOI waveguide. (a) The SOI WG ends in a laterally down-tapered section, whereas the PWB features a section that is laterally and vertically down-tapered to adiabatically transform the SOI waveguide mode to that of the PWB. The PWB waveguide section gradually changes its cross-section from rectangular to circular. (b) The SOI WG tip has a width of $w_{tip} = 80$ nm. The SOI WG taper connects to a straight SOI WG with a cross-section of $w_{Si} = 500$ nm by $h_{Si} = 220$ nm. The lower-index cladding material is not displayed here for the sake of clarity.

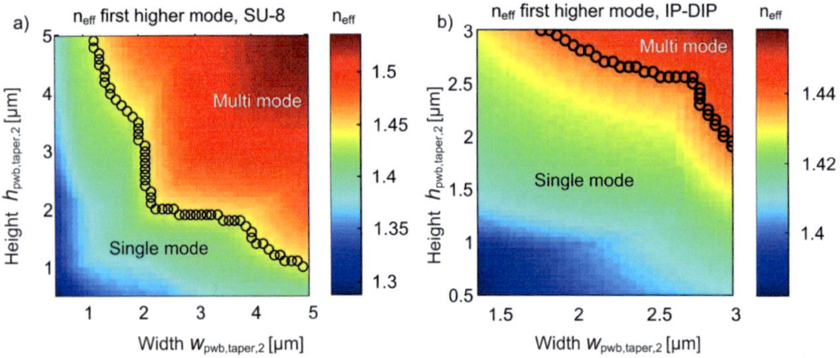

Fig. A.3: Simulated effective refractive index n_{eff} of the first higher mode for (a) a rectangular SU-8 ($n_{PWB,core} = 1.57$ at 1550 nm) waveguide on a SiO_2 ($n_{SiO2} = 1.44$ at 1550 nm) substrate embedded into a lower index cladding material ($n_{clad} = 1.34$ at 1550 nm). The border between single mode and multi-mode dimensions is indicated by black circles. (b) Gives n_{eff} of the first higher mode for a rectangular *IP-Dip*™[39] ($n_{PWB,core} = 1.53$ at 1550 nm) waveguide on a SiO_2 ($n_{SiO2} = 1.44$ at 1550 nm) substrate embedded into a lower index cladding material ($n_{clad} = 1.34$ at 1550 nm). For many experiments a cross-section of 2.8 µm × 2 µm has been used, which still provides single mode waveguiding.

A.5.2 Influence of Photonic Wire Bond Cross-Section

To obtain an insight into the influence of different parameters on the coupling efficiency between the SOI waveguide and the photonic wire bond waveguide section, the ground mode transmission indicated by the S21 parameter of the 3D double-taper (Fig. 3.6) is simulated between the fundamental eigenmode of the straight SOI WG and of the clad rectangular polymer taper WG. A finite integration technique (FIT) solver [61] is implemented, and the following additional data, all valid at a wavelength of $\lambda = 1550$ nm, are used: Refractive indices of silicon ($n_{Si} = 3.48$), silicon dioxide ($n_{SiO2} = 1.44$), PWB taper ($n_{PWB,core} = 1.53$), and cladding ($n_{PWB,clad} = 1.34$). As a first parameter set the influence of the PWB tapers width $w_{PWB,taper,2}$ and height $h_{PWB,taper,2}$ have been studied for an *IP-Dip*™[39] ($n_{PWB,core} = 1.53$) waveguide embedded into a lower refractive index material ($n_{PWB,clad} = 1.34$). In the simulations the SOI taper tip was set fixed to $w_{tip} = 80$nm, the PWB taper tip was set fixed to $w_{PWB,taper,1} = 0.76$ µm and $h_{PWB,taper,1} = 0.45$ µm. The length of the polymer taper amounts to $l_{PWB,taper} = 80$ µm and the length of the SOI taper $l_{Si} = 60$ µm. The SOI WG taper is connected to a straight SOI WG with a cross-section of $w_{Si} = 500$ nm (width) by $h_{Si} = 220$ nm (height). As can be seen in Fig. A.4, the optimum coupling is reached at a PWB waveguide width $w_{PWB,taper,2} = 3.8$ µm and a waveguide height of $h_{PWB,taper,2} = 1.6$ µm. Optimum ground mode transmission in this case amounts to 0.65 dB. Since these dimensions excite the local single-mode conditions, smaller coupling dimensions are selected in most experimental cases, trading in coupling efficiency in order to avoid for mode conversion loss. The simulation shows that the area of 0.1 dB additional coupling loss is relatively broad which is favorable in terms of fabrication tolerances.

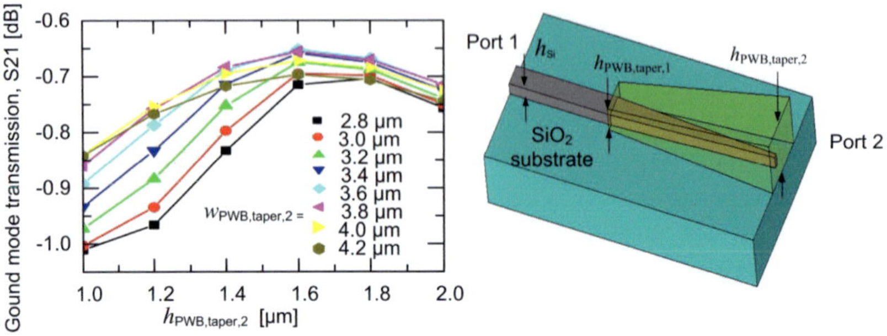

Fig. A.4: Simulated ground mode transmission S21 between the ground mode of an SOI waveguide (Port 1) and the PWB waveguide (Port 2). The following parameters are used for the simulation: SOI parameters: SOI taper tip $w_{\mathrm{tip}} = 80$nm, length of the SOI taper $l_{\mathrm{Si}} = 60$ µm, PWB taper tip $w_{\mathrm{PWB,taper,1}} = 0.76$ µm and $h_{\mathrm{PWB,taper,1}} = 0.45$ µm. length of the polymer taper $l_{\mathrm{PWB,taper}} = 80$ µm. $n_{\mathrm{PWB,core}} = 1.53$, $n_{\mathrm{SiO2}} = 1.44$, $n_{\mathrm{clad}} = 1.34$ all refractive indices are given at lambda=1550 nm. Optimum coupling is reached at a width $w_{\mathrm{PWB,taper,2}} = 3.8$ µm and $h_{\mathrm{PWB,taper,2}} = 1.6$ µm. Since these dimensions excite the local single mode dimensions of the taper smaller dimensions are usually used in the experiment.

A.5.3 Influence of the Silicon Taper-Width

The next influencing factor to be discussed is the silicon taper tip width w_{tip}. The single parameters are visualized in Fig A.2. The length of the SOI taper is set to $l_{\mathrm{Si}} = 60$ µm, the PWB taper is kept at $w_{\mathrm{PWB,taper,1}} = 0.76$ µm and $h_{\mathrm{PWB,taper,1}} = 0.45$ µm, and the length of the two tapers are fixed at $l_{\mathrm{PWB,taper}} = 80$ µm and $l_{\mathrm{Si}} = 60$ µm ($n_{\mathrm{PWB,core}} = 1.53$, $n_{\mathrm{SiO2}} = 1.44$, $n_{\mathrm{clad}} = 1.34$). The silicon taper tip w_{tip}, the width $w_{\mathrm{PWB,taper,2}}$ and height $h_{\mathrm{PWB,taper,2}}$ of the PWB taper are varied. The optimum resulting transmission over SOI taper tip w_{tip} is depicted in Fig A.5.

While a change of the PWB taper width and height of several hundred nanometers only results in additional losses of less than 0.1 dB (see Fig.A.4) the influence of the silicon taper width w_{tip} is much more pronounced. With increasing taper tip width the losses of the coupling efficiency of the 3D double taper structure is tremendously increased. Whereas at SOI tapers with a tip width of $w_{\mathrm{tip}} = 80$ nm a good transmission of -0.65 dB is reached, the optimum transmission droppes to -1.8 dB at an SOI taper tip widths of $w_{\mathrm{tip}} = 200$ nm.

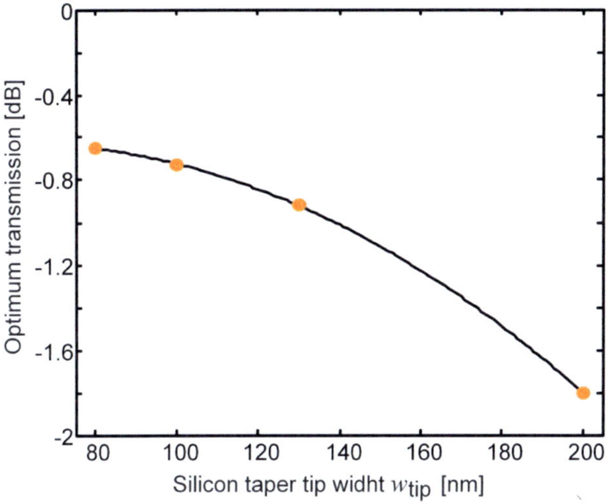

Fig. A.5: Optimum transmission over SOI taper tip width w_{tip}. The following parameters have been used for simulation: while the length of the SOI taper l_{Si} = 60 µm, PWB taper width $w_{\text{PWB,taper,1}}$ = 0.76 µm and height $w_{\text{PWB,taper,1}}$ = 0.45 µm, length of the two tapers $l_{\text{PWB,taper}}$ = 80 µm, l_{Si} = 60 µm. Refractive indices: n_{PWB} = 1.53, n_{SiO2} = 1.44, n_{clad}=1.34 all given at λ = 1550 nm. The silicon taper tip w_{tip}, the width $w_{\text{PWB,taper,2}}$ and height $h_{\text{PWB,taper,2}}$ of the PWB taper have been varied and the optimum transmission is given over SOI taper tip width. As it is clearly visible the SOI taper tip width w_{Si} is a crucial influence factor on the overall transmission of the 3D double taper structure.

It is therefore crucial to use smallest possible taper tip width in order to obtain reasonable optical transmission through the 3D double taper structure. Due to practical reasons, however, taper tip width smaller than w_{tip} = 80 nm are not available from foundry services at the present stage.

A.5.4 Influence of the Silicon Taper-Length

After having identified the SOI taper tip width w_{tip} as a crucial influence factor the influence of the SOI taper length is investigated with the following setup: the SOI WG tip has a width of w_{tip} = 80 nm. Measured from this tip, the investigated SOI WG taper lengths lie between l_{Si} = 40 µm and l_{Si} = 100 µm. The SOI WG taper connects to a straight SOI WG with a cross-section of

w_{Si} = 500 nm (width) by h_{Si} = 220 nm (height). The PWB taper has a cross-section of $w_{PWB,taper,1}$ = 0.76 µm by $h_{PWB,taper,1}$ = 0.45 µm and ends with a cross-section of $w_{PWB,taper,2}$ = 2 µm by $h_{PWB,taper,2}$ = 2 µm at the PWB waveguide side which represents a non-optimized case. The length of the polymer taper is fixed to $l_{PWB,taper}$ = 80 µm.

Except for the l_{Si} = 40 µm and l_{Si} = 100 µm long SOI WG taper, the calculated losses lie between 0.77 dB and 0.9 dB in the C-Band (1530 nm – 1565 nm), see Fig. A.6. Obviously, the 40 µm taper is too short for a low-loss transition. The length of the SOI taper l_{si} is also not a crucial influence factor and a broad range of different length results in acceptable transmission.

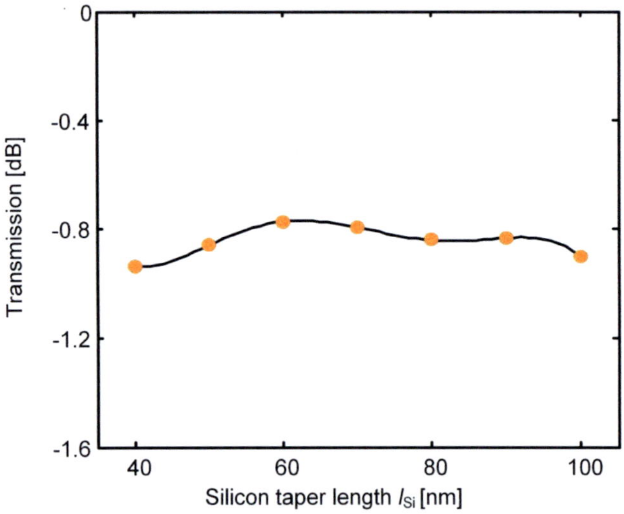

Fig. A.6: Simulated transmission versus silicon taper length at lambda=1550 nm for a 3D double-taper interface as shown in Fig. 3.6. The following values apply for the SOI waveguide: w_{Si} = 500 nm, h_{Si} = 220 nm, taper tip width w_{tip} = 80 nm, refractive index n_{Si} = 3.48 at 1550 nm. For the PWB taper: initial height $h_{PWB,taper,1}$ = 450 nm, initial width $w_{PWB,taper,1}$ = 760 nm, PWB waveguide section $w_{PWB,taper,2}$ = 2 µm × $h_{PWB,taper,2}$ = 2 µm, $n_{PWB,core}$ = 1.53 and $n_{PWB,clad}$ = 1.34 both at 1550 nm. The silicon taper length l_{Si} is varied in the simulation between l_{Si} = 40 µm and l_{Si} = 100 µm. Except the l_{Si} = 40 µm and l_{Si} = 100 µm long SOI taper the simulation losses range between 0.75 dB and 0.9 dB.

A.5.5 Influence of the Photonic Wire Bond Taper Length

The influence of the SOI taper length has been studied in the last paragraph, now the other parameters are kept constant and the PWB taper length $l_{\mathrm{PWB,taper}}$ is varied. The PWB taper has a cross-section of $w_{\mathrm{PWB,taper,1}} = 0.76$ µm by $h_{\mathrm{PWB,taper,1}} = 0.45$ µm and ends with a cross-section of $w_{\mathrm{PWB,taper,2}} = 2.4$ µm by $h_{\mathrm{PWB,taper,2}} = 1.6$ µm at the PWB waveguide side. The SOI taper tip has a width w_{si}= 100 nm. The length of the polymer taper is fixed to l_{Si}= 60 µm. The PWB taper length varies between $l_{\mathrm{PWB,taper}} = 60$ µm and $l_{\mathrm{Si}} = 100$ µm. Again *IP-Dip*™[39] with a refractive index of 1.53 is chosen as a waveguide material, located on top of a SiO_2 substrate ($n_{SiO2} = 1.44$) embedded into a lower index cladding material (n_{clad}= 1.34). As can be seen in Fig. A.7 the influence of the PWB taper length is small and a broad range of different taper length does not lead to significant changes of the transmission of the double taper structure.

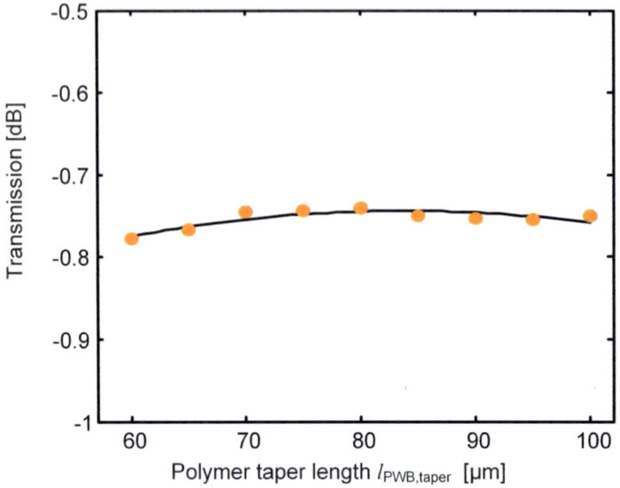

Fig. A.7: Simulated transmission versus PWB taper length at lambda=1550 nm for a 3D double-taper interface as shown in Fig. 3.6. The following values apply for the SOI waveguide: $w_{\mathrm{Si}} = 500$ nm, $h_{\mathrm{Si}} = 220$ nm, taper tip width $w_{\mathrm{tip}} = 100$ nm, refractive index $n_{\mathrm{Si}} = 3.48$ at 1550 nm. For the PWB taper: initial height $h_{\mathrm{PWB,taper,1}} = 450$ nm, initial width $w_{\mathrm{PWB,taper,1}} = 760$ nm, PWB waveguide section $w_{\mathrm{PWB,taper,2}} = 2.4$ µm $\times$ $h_{\mathrm{PWB,taper,2}} = 1.8$ µm, $n_{\mathrm{PWB,core}} = 1.53$ and $n_{\mathrm{clad}} = 1.34$ both at 1550 nm. The silicon taper length $l_{\mathrm{Si}} = 60$ µm and the $l_{\mathrm{PWB,taper}}$ is varied between 60 µm and 100 µm. The transmission is almost independent of the chosen PWB taper length.

A.5.6 Cladding Refractive Index

The efficiency of the double taper mode size converter depends on the used cladding material. The influence of the cladding material is studied by changing the refractive index of the background material and simulating the optical transmission between the PWB waveguide section and the un-tapered SOI waveguide. The following simulation values are chosen:

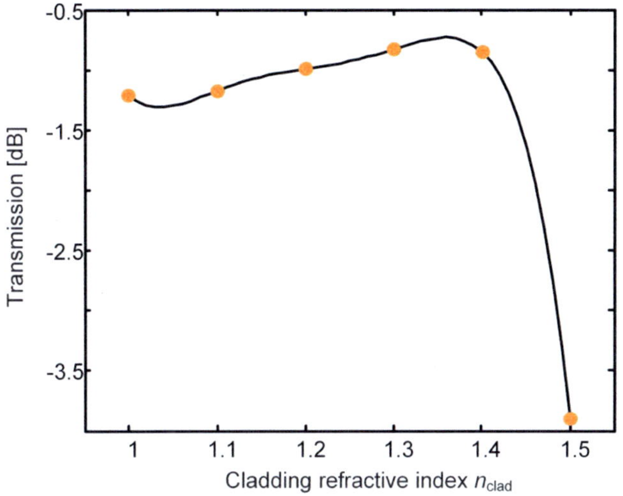

Fig. A.8: Simulated transmission versus cladding refractive index n_{clad} at lambda=1550 nm for a 3D double-taper interface as shown in Fig. 3.6. The following values apply for the SOI waveguide: w_{Si} = 500 nm, h_{Si} = 220 nm, taper tip width w_{tip} = 100 nm, refractive index n_{Si} = 3.48 at 1550 nm. For the PWB taper: initial height $h_{PWB,taper,1}$ = 450 nm, initial width $w_{PWB,taper,1}$ = 760 nm, PWB waveguide section $w_{PWB,taper,2}$ = 2.4 µm × $h_{PWB,taper,2}$ = 1.8 µm, $n_{PWB,core}$ = 1.53. The silicon taper length l_{Si} = 60 µm and the $l_{PWB,taper}$ = 80 µm . The refractive index of the cladding material is varied between n_{clad} = 1 and $n_{PWB,clad}$ = 1.5. As it is visible in the simulation result, the transmission worsens dramatically once the cladding refractive index goes beyond 1.4. The mode is not well guided in the PWB waveguide anymore, such that an efficient optical transmission is between the eigenmode in the PWB waveguide and the eigenmode of the un-tapered SOI waveguide is no longer possible.

Silicon waveguide width and height: $w_{Si} = 500$ nm, $h_{Si} = 220$ nm, taper tip width $w_{tip} = 100$ nm, refractive index $n_{Si} = 3.48$ at 1550 nm. For the PWB taper: initial height $h_{PWB,taper,1} = 450$ nm, initial width $w_{PWB,taper,1} = 760$ nm, PWB waveguide section $w_{PWB,taper,2} = 2.4$ µm × $h_{PWB,taper,2} = 1.8$ µm, $n_{PWB,core} = 1.53$ and. The silicon taper length $l_{Si} = 60$ µm and the $l_{PWB,taper} = 80$ µm. The refractive index of the cladding material is varied between $n_{clad} = 1$ and $n_{PWB,clad} = 1.5$. As can be seen from Fig. A.8, starting from a cladding refractive index of 1.4 the optical losses tremendously increase. Due to the low refractive index contrast between PWB core and cladding the modal cross-section becomes higher and an efficient coupling between this mode and the SOI waveguide mode is no longer possible with the given parameter set. The cladding material is therefore a crucial influence factor and has to be taken into account when designing the 3D double taper.

A.5.7 Wavelength Dependency of the Coupling Structure

One major advantage of the 3D double taper approach is its broadband transmission characteristic. Unlike grating couplers with their intrinsically limited bandwidth - typical 1 dB bandwidths amount to 40-50 nm - the 3dB double taper approach here proposed provides a much broader 1 dB bandwidth which can be seen in Fig. A.9. Simulation indicates a 1 dB bandwidth much larger than 100 nm. Since the actual dispersion relations of the used polymer materials are unknown and not taken into account in the simulation a detailed evaluation of the broadband behavior of photonic wire bond SOI interconnects is left open for experimental demonstration.

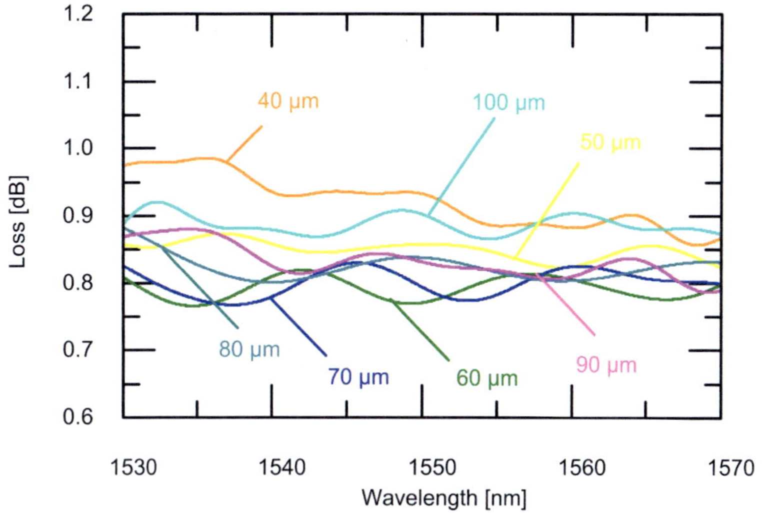

Fig. A.9: Simulated wavelength dependent transmission for a 3D double-taper interface as shown in Fig. 3.6 for different PWB taper length $l_{PWB,taper}$: The following values apply for the SOI waveguide: $w_{Si} = 500$ nm, $h_{Si} = 220$ nm, taper tip width $w_{tip} = 80$ nm, refractive index $n_{Si} = 3.48$ at 1550 nm. For the PWB taper: initial height $h_{PWB,taper,1} = 450$ nm, initial width $w_{PWB,taper,1} = 760$ nm, PWB waveguide section $w_{PWB,taper,2} = 2$ µm × $h_{PWB,taper,2} = 2$ µm, $n_{PWB,core} = 1.53$ and $n_{PWB,clad} = 1.34$ both at 1550 nm. Since the dispersion relations of the used polymer materials are unknown the evaluation of a broader wavelength range is left for experimental investigation.

A.6 Fabrication Details [J1]

Table A.4: Fabrication Details for the PWB sample shown in Chapter 5.7 and published in [J1]

Substrate: SOI chip	
Width SOI WG:	w_{si} = 500 nm
Height SOI WG:	h_{si} = 220 nm
Mounting:	Glass carrier with two height levels
Coupling structures on chip: Waveguide taper	
Taper length:	l_{si} = 60 µm
Taper tip width:	w_{tip} = 80 nm
Material and sample preparation:	
Chip mounting:	Dymax OP-4-20632
PWB material:	*IP-Dip*™ [38]
Refractive index core:	n_{co} = 1.53 at 1550 nm
Cladding:	Index Oil
Refractive index cladding:	n_{clad} = 1.34 at 1550 nm
Cleaning:	Bath in aceton Bath in isopropyl alcohol Dried in nitrogen flow
Developer:	PGMEA
Setup:	
Sample setup:	Dip-in configuration, see Fig. 1.4
Lithography setup:	Piezo-based see Section 1.3.3

On-chip positioning: Port detection algorithm on SOI chip see Section 2.1. manual detection of fiber cores		
Writing strategy: Interior: Helix writing, see Section 2.4.2 Shell: Trajectory writing see Section 2.4.1		
Distance between lines:	100 nm in all directions	
Velocity:	75 µm/s	
PWB parameters:		
Trajectory:	5th order polynomial, trajectory planning algorithm	
PWB width:	$w_{\mathrm{PWB,taper,2}} = 2\ \mu\mathrm{m}$	
PWB height:	$h_{\mathrm{PWB,taper,2}} = 2\ \mu\mathrm{m}$	
Fiber - side:	PWB taper tip:	$d_{\mathrm{taper,2}} = 12\ \mu\mathrm{m}$
	PWB taper length:	$l_{\mathrm{up\text{-}taper}} = 60\ \mu\mathrm{m}$
SOI - side:	PWB taper tip width:	$w_{\mathrm{PWB,taper,1}} = 0.76\ \mu\mathrm{m}$
	PWB taper tip height:	$h_{\mathrm{PWB,taper,1}} = 0.45\ \mu\mathrm{m}$
	PWB taper length:	$L_{\mathrm{PWB,taper}} = 80\ \mu\mathrm{m}$

A.7 Calculation and simulation of field distributions in the focal region

In the following several details about the derivation of the electrical field distribution in the focal region as well as on details of the simulations performed Chapter 1 and Chapter 0 are given.

Maxwell's equations are used in the form [91]:

$$\nabla \cdot \vec{D}(\vec{r},t) = 0 \tag{A.7.1}$$

$$\nabla \times \vec{\mathcal{E}}(\vec{r},t) = -\frac{\partial \vec{B}(\vec{r},t)}{\partial t} \tag{A.7.2}$$

$$\nabla \cdot \vec{B}(\vec{r},t) = 0 \tag{A.7.3}$$

$$\nabla \times \vec{\mathcal{H}}(\vec{r},t) = \frac{\partial \vec{D}(\vec{r},t)}{\partial t} \tag{A.7-4}$$

with the magnetic flux density $\vec{B}$

$$\vec{B}(\vec{r},t) = \mu_0 \mathcal{H}(\vec{r},t) \tag{A.7.5}$$

and the electric field $\vec{\mathcal{E}}$, the dielectric displacement $\vec{D}$, and the polarization $\vec{P}$

$$\vec{D}(\vec{r},t) = \epsilon_0 \vec{\mathcal{E}}(\vec{r},t) + \vec{P}(\vec{r},t) \tag{A.7.6}$$

The mathematical description presented in this section has been developed in close collaboration with Maria-Laura Gödecke [74]. The following section is a quotation from her master thesis.[1]

– [Beginning quotation from [74]] –

[1] The images in this sections have been reformatted, reference numbering and figure numbering have been changed for consistency.

A.7.1 The electromagnetic field in the focal region

When focusing a laser beam under large opening angles, the field distribution in the focal region significantly deviates from the simple form of a Gaussian beam. The reason is straightforward: The analytic formula describing a Gaussian beam is based on the paraxial approximation, which is valid for beams which are focused under half opening angles smaller than 15° [92]. In our experimental setup, the half opening angle amounts to approximately 70°, which by far exceeds the limiting value mentioned before.

In this section, we will derive an integral representation for the electromagnetic field distribution in the focal region of an optical imaging system, which is valid for systems of low as well as high numerical aperture. We will largely follow the analysis of Stamnes [75]. Additional sources will be quoted in the main text.

A.7.2 Wave propagation and diffraction

For our analysis, we adopt Cartesian coordinates $q = x, y, z$. We consider a linear, homogeneous, lossless and time-invariant medium with unity relative permeability. Furthermore, we assume monochromatic fields of angular frequency ω, whose time dependence is $\mathrm{e}^{-\mathrm{i}\omega t}$. The electric field vector $\vec{\mathcal{E}}(\vec{r},t)$ at position $\vec{r} = (x, y, z)$ and time t is given by

$$\vec{\mathcal{E}}(\vec{r},t) = \mathrm{Re}\{\vec{E}(\vec{r})\mathrm{e}^{-\mathrm{i}\omega t}\}. \tag{A.7.7}$$

The vector $\vec{E}$ which is to be determined is generally complex-valued. The magnetic field vector $\vec{\mathcal{H}}(\vec{r},t)$ is defined analogously:

$$\vec{\mathcal{H}}(\vec{r},t) = \mathrm{Re}\{\vec{H}(\vec{r})\mathrm{e}^{-\mathrm{i}\omega t}\}. \tag{A.7.8}$$

In a region free of sources, each component E_q of $\vec{E}$ satisfies the scalar Helmholtz equation

$$\left(\nabla^2 + k^2\right)E_q\left(\vec{r}\right) = 0, \tag{A.7.9}$$

where $k = 2\pi n / \lambda$ is the propagation constant in the medium of refractive index n. The medium wavelength is denoted by λ / n.

We may readily verify that a plane wave $e^{i\vec{k}\vec{r}}$ satisfies Eq. (A.7.9). By superposition, a general solution is therefore of the form

$$\vec{E}(\vec{r}) = \left(\frac{1}{2\pi}\right)^2 \iint_{-\infty}^{+\infty} \vec{E}\left(k_x, k_y\right) e^{i\vec{k}\cdot\vec{r}} dk_x dk_y. \qquad \text{(A.7.10)}$$

Eq. (2.5) is called an angular spectrum of plane waves, with the angular spectrum $\vec{E}(k_x, k_y)$ being the complex amplitude of the plane wave with propagation vector $\vec{k} = k_x \vec{e}_x + k_y \vec{e}_y + k_z \vec{e}_z$. $\vec{e}_x$, $\vec{e}_y$ and $\vec{e}_z$ denote the unit vectors in x-, y- and z-direction, respectively. We restrict our attention to fields that propagate in the positive z-direction. Hence, k_z is given by

$$k_z = \begin{cases} +\left(k^2 - k_x^2 - k_y^2\right)^{1/2} & \text{if} \quad k_x^2 + k_y^2 \le k^2, \\ +i\left(k_x^2 + k_y^2 - k^2\right)^{1/2} & \text{if} \quad k_x^2 + k_y^2 > k^2. \end{cases} \qquad \text{(A.7.11)}$$

When $k_x^2 + k_y^2 \le k^2$, the corresponding plane waves are of the homogeneous type propagating in the positive z-direction, but when $k_x^2 + k_y^2 > k^2$, they are of the inhomogeneous or evanescent type decaying exponentially with increasing z. $\mathrm{Im}\{k_z\} > 0$ ensures that the radiation condition is fulfilled.

Let us now assume that the field $\vec{E}(\vec{r})$ is radiated by sources in the half-space $z > 0$ and that we know the field $\vec{E}(x, y, 0)$ in the boundary plane $z = 0$. Its spatial Fourier transform is the angular spectrum $\vec{E}\left(k_x, k_y\right)$:

$$\vec{E}\left(k_x, k_y\right) = \iint_{-\infty}^{+\infty} \vec{E}(x, y, 0) e^{-i\left(k_x x + k_y y\right)} dx dy. \qquad \text{(A.7.12)}$$

The field at an arbitrary position in the half-space $z > 0$ may be obtained from the angular-spectrum representation in Eq. (A.7.10) together with the angular spectrum given by Eq. (A.7.12).

Eq. (A.7.10) and (A.7.12) constitute the first Rayleigh-Sommerfeld integral (Dirichlet boundary conditions), written as a plane-wave expansion [93]. If we knew the normal derivative $\partial\vec{E}(x,y,z)/\partial z$ of the field in the boundary plane $z=0$ instead of the field $\vec{E}(x,y,0)$ itself, we would have to use the second Rayleigh-Sommerfeld integral (Neumann boundary conditions) [93].

One significant advantage of the angular-spectrum representation in Eq. (A.7.10) over the corresponding impulse-response integral is that both the angular spectrum $\hat{\vec{E}}(k_x,k_y)$ and the field $\vec{E}(\vec{r})$ in planes $z=\text{constant}$ can be efficiently computed by means of a fast Fourier Transform (FFT) algorithm. We will return to this point in more detail later on.

It should be emphasized that the solution presented in Eq. (A.7.12) and (A.7.10) is exact: So far, we have made no approximations. However, the exact specification of the values of the field $\vec{E}(x,y,0)$ in the boundary plane $z=0$ presents formidable difficulties. For this reason, approximate solutions are usually desired.

Let us assume that the incident field is diffracted through an aperture in a perfectly-absorbing plane screen at $z=0$. One common approximation to the boundary values is the so-called Kirchhoff approximation. It implies that both the field and its normal derivative inside the aperture are assumed to be equal to the values they would have there if the screen were not present. At all points outside the aperture, the field and its normal derivative are assumed to vanish.

Before procceding we point out that outside the aperture, there are discontinuous jumps in the values of the field and its normal derivative from one side of the screen to the other. Hence, a solution based on the Kirchhoff approximation cannot be expected to represent the field correctly at points in the direct vicinity of the aperture. However, when the aperture is large compared to the wavelength and when both the geometrical focus P_1 and the observation point P_2 are located at distances from the aperture that are large compared to the wavelength, we obtain a good approximation to the true field [94].

A.7.3 Focusing of electromagnetic waves

Having derived the necessary formulas in the previous section, we may now turn our attention to the focusing of electromagnetic waves. Fig. A.10 illustrates the studied configuration. A point source at P_0 generates a diverging spherical wave which is imaged by an optical system. The optical axis coincides with the z-axis of the Cartesian coordinate system; the back vertex plane of the last lens is located in the x-y-plane. The image space is filled with a single homogeneous medium. The transmitted wave features an aberrated converging wavefront with its geometrical focus at the point P_1 with position vector $\vec{r}_1 = (x_1, y_1, z_1)$.

Using the Kirchhoff approximation, the field in the plane $z = 0$ is given by

$$\vec{E}(x,y,0) = \begin{cases} \left[\vec{E}_0(x,y)/R_1\right] \exp\left[\mathrm{i}k\phi_0(x,y) - \mathrm{i}kR_1\right] & \text{inside } \mathcal{A}, \\ 0 & \text{outside } \mathcal{A}, \end{cases} \qquad \text{(A.7.13)}$$

where

$$R_1(x,y) = \left[(x-x_1)^2 + (y-y_1)^2 + z_1^2)\right]^{1/2}. \qquad \text{(A.7.14)}$$

$\mathcal{A}$ denotes the aperture domain. The vectorial amplitude function $\vec{E}_0(x,y)()$ can be derived from geometrical ray tracing data, taking into account the polarization state of the field along each ray. The aberration function $\phi_0(x,y)$ represents the phase deviation from the (non-aberrated) converging spherical wavefront. It may likewise be determined from the ray trace.

We observe the diffracted field at an arbitrary point P_2 with position vector $\vec{r}_2 = (x_2, y_2, z_2)$ in the half-space $z > 0$. Against this backdrop, Eq. (A.7.10) reads

$$\vec{E}(\vec{r}_2) = \left(\frac{1}{2\pi}\right)^2 \iint_{-\infty}^{+\infty} \vec{E}(k_x, k_y) \mathrm{e}^{\mathrm{i}\vec{k}\cdot\vec{r}_2} \mathrm{d}k_x \mathrm{d}k_y. \qquad \text{(A.7.15)}$$

For simplicity, we assume an aberration-free imaging system: $\phi_0(x,y)=0$. Substituting Eq. (A.7.13) into Eq. (A.7.12) yields

$$\vec{E}\left(k_x,k_y\right)=\iint_{\mathcal{A}}\frac{\vec{E}_0\left(x,y\right)}{R_1}\mathrm{e}^{-\mathrm{i}kR_1}\mathrm{e}^{-\mathrm{i}\left(k_x x+k_y y\right)}\mathrm{d}x\mathrm{d}y. \tag{A.7.16}$$

The integrands in Eq. (A.7.15) and (A.7.16) usually vary rapidly over the integration domain, which is why the numerical evaluation of these double integrals is time-consuming. Computational savings can be gained by employing asymptotic evaluation techniques.

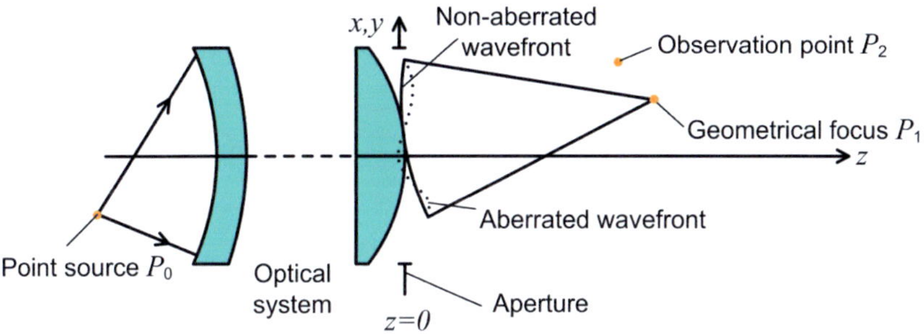

Fig. A.10: Imaging of a point source by an optical system [75]. A diverging spherical wavefront, generated by a point source at P_0, is transformed into an aberrated converging wavefront with its geometrical focus at P_1. The field is observed at an arbitrary point P_2 in the half-space $z>0$

We recall that the focal point P_1 is located at a distance from the aperture plane that is large compared to the wavelength: $kz_1 \gg 1$. Under these circumstances, the angular-spectrum integral in Eq. (A.7.16) may be evaluated asymptotically by applying the method of stationary phase. The corresponding formulas are derived in Appendix A.1[2]. Using the asymptotic approximation instead of the exact value introduces an error on the order of $\left(kR_1\right)^{-3/2}$ [94].

For notational convenience, we write

[2] The Appendix A.1 of [74] is no part of this quotation.

$$\vec{g}(x,y)=\frac{\vec{E}_0(x,y)}{R_1}, \tag{7.1}$$

$$f(x,y)=-\frac{kR_1+k_x x+k_y y}{k}, \tag{A.7.18}$$

which transforms Eq. (A.7.16) into

$$\vec{E}(k_x,k_y)=\iint_{\mathcal{A}}\vec{g}(x,y)\mathrm{e}^{\mathrm{i}kf(x,y)}\mathrm{d}x\mathrm{d}y. \tag{A.7.19}$$

The phase function $f(x,y)$ is stationary, i.e. $f_x(x,y)=f_y(x,y)=0$, at the stationary point (x_s,y_s), with

$$x_s=x_1-z_1\frac{k_x}{k_z}, \tag{A.7.20}$$

$$y_s=y_1-z_1\frac{k_y}{k_z}. \tag{A.7.21}$$

According to Eq. (A.7.9)[3], the contribution $\vec{E}_s(k_x,k_y)$ of the interior stationary point is (to the lowest order) given by

$$\begin{gathered}\vec{E}_s(k_x,k_y)\approx\frac{-2\pi\mathrm{i}\sigma}{\sqrt{|\alpha\beta-\gamma^2|k^2}}\,\vec{g}(x_s,y_s)\mathrm{e}^{\mathrm{i}kf(x_s,y_s)},\\ \begin{aligned}\alpha&=f_{xx}(x_s,y_s),\\ \beta&=f_{yy}(x_s,y_s),\\ \gamma&=f_{xy}(x_s,y_s),\end{aligned}\ \sigma=\begin{cases}+1 & \text{if}\quad \alpha\beta>\gamma^2 \quad\text{and}\quad \alpha>0,\\ -1 & \text{if}\quad \alpha\beta>\gamma^2 \quad\text{and}\quad \alpha<0,\\ +\mathrm{i} & \text{if}\quad \alpha\beta<\gamma^2.\end{cases}\end{gathered} \tag{A.7.22}$$

---

[3] Equation A.7.9:

$$\begin{gathered}\int_A g(x,y)\mathrm{e}^{\mathrm{i}kf(x,y)}\mathrm{d}x\mathrm{d}y\approx\frac{-2\pi\mathrm{i}\sigma}{\sqrt{|\alpha\beta-\gamma^2|k^2}}g(x_s,y_s)\mathrm{e}^{\mathrm{i}kf(x_s,y_s)},\\ \begin{aligned}\alpha&=f_{xx}(x_s,y_s),\\ \beta&=f_{yy}(x_s,y_s),\\ \gamma&=f_{xy}(x_s,y_s),\end{aligned}\ \sigma=\begin{cases}+1 & \text{if } \alpha\beta>\gamma^2 \text{ and } \alpha>0,\\ -1 & \text{if } \alpha\beta>\gamma^2 \text{ and } \alpha<0,\\ +\mathrm{i} & \text{if } \alpha\beta<\gamma^2.\end{cases}\end{gathered}$$

By direct calculation, we readily find that $\alpha<0$, $\beta<0$ and $\alpha\beta>\gamma^2$, so that $\sigma=-1$. Furthermore,

$$\alpha\beta-\gamma^2=\frac{k_z^4}{z_1^2k^4}, \tag{A.7.23}$$

$$\vec{g}(x_s,y_s)=\frac{\vec{E}_0(x_s,y_s)}{R_1(x_s,y_s)}=\frac{k_z}{z_1k}\vec{E}_0(x_s,y_s), \tag{A.7.24}$$

$$f(x_s,y_s)=-\frac{\vec{k}\cdot\vec{r}_1}{k}. \tag{A.7.25}$$

With these relations, Eq. A.7.22 becomes

$$\vec{E}_s(k_x,k_y)=\begin{cases}-(2\pi\mathrm{i}/k_z)\vec{E}_0(x_s,y_s)\mathrm{e}^{-\mathrm{i}\vec{k}\cdot\vec{r}_1} & \text{inside } \Omega_1,\\ 0 & \text{outside } \Omega_1,\end{cases} \tag{A.7.26}$$

where Ω_1 is the solid angle subtended by the aperture at the focal point P_1. The case differentiation results from the requirement that the interior stationary point shall be located well inside the aperture domain, i.e. $(x_s,y_s)\in\mathcal{A}$.

Fig. A.11 shows the studied configuration from a different perspective, thus illustrating the case differentiation mentioned before. We assume a circular aperture and draw straight lines from the focal point P_1 through the edge of the aperture, which results in a geometrical cone. Only those plane waves whose propagation vectors reach the image space through the aperture, that is to say those whose propagation vectors fall inside the cone, contribute to the field in the focal region. Hence, there is an abrupt cut-off of the angular-spectrum of the incident field.

According to Eq. (A.7.16), $\vec{E}(k_x,k_y)$ is obtained by adding the contributions of an infinite number of secondary waves, one from each point in the aperture. The asymptotic evaluation of this angular-spectrum integral results in a simpler expression, with only a few secondary waves contributing significantly to $\vec{E}_s(k_x,k_y)$, compare with Eq. (A.7.26). In fact, no integration remains at all, which implies a tremendous saving in computing time.

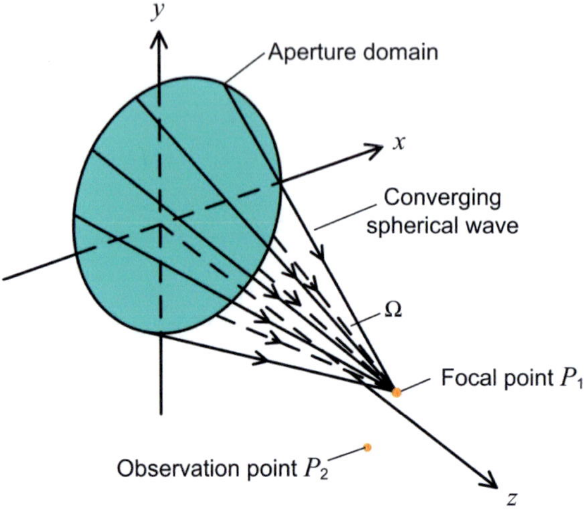

Fig. A.11: Schematic overview of the studied configuration from a different perspective [75]. Drawing straight lines from the focal point P_1 through the edge of the circular aperture creates a geometrical cone.

A commonly used approximation to the angular spectrum of a focused field is the so-called Debye approximation [95]. It consists of retaining only the dominant contribution of the interior stationary point [96], i.e. we set

$$\vec{E}\left(k_x,k_y\right)=\vec{E}_S\left(k_x,k_y\right). \tag{A.7.27}$$

We substitute Eq. (A.7.27) and Eq. (A.7.26) into Eq. (A.7.15) and obtain the Debye integral representation for a non-aberrated focused field:

$$\vec{E}\left(\vec{r}_2\right)=\frac{-\mathrm{i}}{2\pi}\iint_{\Omega_1}\frac{\vec{E}_0\left(x_s,y_s\right)}{k_z}\mathrm{e}^{\mathrm{i}\vec{k}\left(\vec{r}_2-\vec{r}_1\right)}\mathrm{d}k_x\mathrm{d}k_y. \tag{A.7.28}$$

Finally, we use the notation

$$\vec{E}_0\left(x_s,y_s\right)=\vec{a}\left(k_x,k_y\right) \tag{A.7.29}$$

and shift the origin of our Cartesian coordinate system to the focal point P_1, i.e. we write $\vec{r} = \vec{r}_2 - \vec{r}_1$, where $\vec{r}$ is the radius vector from the geometrical focal point P_1 to the observation point P_2. Eq. (A.7.28) then reads

$$\vec{E}(\vec{r}) = -\frac{\mathrm{i}}{2\pi} \iint_{\Omega_1} \frac{\vec{a}(k_x, k_y)}{k_z} e^{\mathrm{i}\vec{k}\cdot\vec{r}} \mathrm{d}k_x \mathrm{d}k_y. \tag{A.7.30}$$

An analogous expression for the time-independent part $\vec{H}$ of the magnetic field vector may be derived in an exactly similar manner:

$$\vec{H}(\vec{r}) = -\frac{\mathrm{i}}{2\pi} \iint_{\Omega_1} \frac{\vec{b}(k_x, k_y)}{k_z} e^{\mathrm{i}\vec{k}\cdot\vec{r}} \mathrm{d}k_x \mathrm{d}k_y. \tag{A.7.31}$$

According to Maxwell's equations, the vectorial amplitude functions $\vec{a}(k_x, k_y)$ and $\vec{b}(k_x, k_y)$ obey the relation

$$\vec{b}(k_x, k_y) = \left(\frac{c}{\omega}\right) \vec{k} \times \vec{a}(k_x, k_y). \tag{A.7.32}$$

A.7.4 Validity of the Debye approximation

It can easily be shown that the amplitude of the focused field in Eq. (A.7.30) has inversion symmetry about the focal point, whereas the phase has inversion antisymmetry about the same point. These symmetry relations result from the application of the Debye approximation. However, they are inconsistent with an asymmetry of the focused field together with a focal shift that may be experimentally observed when focusing at low Fresnel numbers (see [97] and the references therein). Since the Debye approximation fails to account for this asymmetry, we should investigate the conditions for the validity of the corresponding integral representation of the focused field.

The discrepancies between experiment and theory arise from the fact that the Debye approximation neglects diffraction at the edge of the aperture. An asymptotic evaluation of the angular-spectrum integral given in Eq. (A.7.16) by the method of stationary phase yields two contributions: one from interior stationary points and one from the edge of the aperture. The first contribution, which is stated in Eq. (A.7.26), constitutes the Debye approximation. It drops abruptly to zero for planes waves whose propagation vectors lie outside the cone depicted in Fig. A.11. The contribution from diffraction at the edge of the aperture, which includes plane waves with propagation vectors outside the cone, is neglected in the Debye approximation.

Wolf and Li [98] established a simple sufficiency condition under which the Debye integral representation of the focused field in Eq. (A.7.30) may be expected to give a good approximation to the true field in the focal region:

$$kR \gg \frac{\pi}{\sin^2\left(\alpha_{\max}/2\right)}. \tag{A.7.33}$$

Here, R is the focal length of the system and $\alpha_{\max}$ is the angular semi-aperture of the focusing geometry, see Fig. A.12. Thus, for a fixed focal length, the Debye approximation gets better the larger the angular aperture; and for a fixed angular aperture, it gets better the larger the focal length. With regard to the parameters of our experimental setup, the condition in Eq. (A.7.33) is usually satisfied.

When the angular semi-aperture is much smaller than 45°, the condition stated in Eq. (A.7.33) reduces to the requirement that the Fresnel number of the focusing geometry is large compared to unity. Hence, focusing at low Fresnel numbers implies deviations from the results obtained by applying the Debye approximation.

A.7.5 Derivation of the vectorial amplitude function

In the following, we will derive an expression for the vectorial amplitude function $\vec{a}\left(k_x,k_y\right)$, which points along the electric field vector of the plane wave with propagation vector $\vec{k}$. Our analysis largely follows the work of van de Nes [76], [77].

We consider a rotationally symmetric optical system which gives rise to an aplanatic projection (i.e. the imaging is axially stigmatic and obeys Abbe's sine condition). Furthermore, we assume the imaging system to be free of absorption or reflection losses. Fig. A.12 depicts an exemplary optical path through the imaging system. A monochromatic plane wavefront is incident along the optical axis pointing in z-direction. We denote the electric field distribution in the entrance pupil by $\vec{E}_\mathrm{i}\left(k_x,k_y\right)$. In the simplest case, the field features a spatially constant amplitude, but we may as well define more general amplitude distributions. Against this backdrop, it is easily possible to study the focusing of Gaussian laser beams.

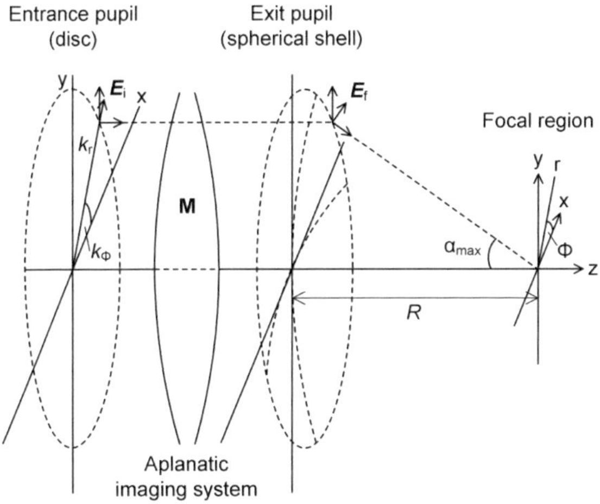

Fig. A.12: Optical path through an aplanatic imaging system [97]. The emergent ray meets the exit pupil at the same height at which the corresponding ray in the object space entered the system. The matrix **M** maps the electric field distribution defined in the entrance pupil onto the exit pupil.

In addition to the Cartesian basis $(\vec{e}_x,\vec{e}_y,\vec{e}_z)$, we introduce a cylindrical basis $(\vec{e}_r,\vec{e}_\phi,\vec{e}_z)$. The transition from the entrance pupil to the exit pupil can be considered as a rotation of the field vector $\vec{E}_\mathrm{i}$ around the angular axis $\vec{e}_\phi$ towards the optical axis. This lens operation is described by the propagation matrix $\mathbf{M}$.

Scalar multiplication of the lens operator with the electric field distribution $\vec{E}_\mathrm{i}\left(k_x,k_y\right)$ in the entrance pupil yields the corresponding distribution in the exit pupil: $\vec{E}_\mathrm{f}\left(k_x,k_y\right)=\mathbf{M}\vec{E}_\mathrm{i}\left(k_x,k_y\right)$.

The propagation associated with $\mathbf{M}$ may be divided into three steps: First, we use the matrix $\mathbf{P}$ to project the Cartesian basis onto the cylindrical basis. Second, we perform the rotation of the electric field vectors in the cylindrical basis by means of scalar multiplication with the matrix $\mathbf{R}$. Third, we use the matrix $\mathbf{P}^{-1}$, which is the inverse of the first operation, to project the cylindrical basis back onto the Cartesian basis. The entire propagation matrix is obtained from $\mathbf{M}=\mathbf{P}^{-1}\mathbf{RP}$. For notational convenience, we decompose the medium propagation vector into cylindrical coordinates: $\vec{k}=\left(k_r,k_\phi,k_z\right)$, with $0\le k_\phi<2\pi$ and $0\le k_r<k\sin\alpha_\mathrm{max}$. The matrices $\mathbf{P}$, $\mathbf{R}$ and $\mathbf{M}$ are then given by

$$\mathbf{P}=\begin{pmatrix}\cos k_\phi & \sin k_\phi & 0\\ -\sin k_\phi & \cos k_\phi & 0\\ 0 & 0 & 1\end{pmatrix}, \tag{A.7.34}$$

$$\mathbf{R}=\frac{1}{k}\begin{pmatrix}k_z & 0 & k_r\\ 0 & k & 0\\ -k_r & 0 & k_z\end{pmatrix}, \tag{A.7.35}$$

$$\mathbf{M}=\mathbf{P}^{-1}\mathbf{RP}=\begin{pmatrix}k_z\cos^2 k_\phi+k\sin^2 k_\phi & \left(k_z-k\right)\sin k_\phi\cos k_\phi & k_r\cos k_\phi\\ \left(k_z-k\right)\sin k_\phi\cos k_\phi & k_z\sin^2 k_\phi+k\cos^2 k_\phi & k_2\sin k_\phi\\ -k_r\cos k_\phi & -k_r\sin k_\phi & k_z\end{pmatrix}. \tag{A.7.36}$$

Each of the entries in the third column of the propagation matrix $\mathbf{M}$ may as well be set equal to zero, since the incident electric field $\vec{E}_\mathrm{i}\left(k_x,k_y\right)$ does not feature a longitudinal component (in z-direction). Nevertheless, the rotated electric field vectors $\vec{E}_{\mathrm{f}}\left(k_x,k_y\right)$ may have non-zero z-components, see Fig. A.12.

As already mentioned, the optical imaging system obeys Abbe's sine condition, which implies that a ray in the image space meets the exit pupil at the same height h at which the corresponding ray in the object space entered the system:

$$h=R\sin\alpha. \tag{A.7.37}$$

Conservation of energy requires that the electric field energy density l_0^2 within an infinitesimally small annular surface element δS_0 on the disc in the entrance pupil should be equal to the electric field energy density l_1^2 within the corresponding surface element δS_1 on the spherical shell in the exit pupil [99]:

$$l_0^2\delta S_0=l_1^2\delta S_1. \tag{A.7.38}$$

The surface elements are given by

$$\delta S_0=2\pi h\mathrm{d}h, \tag{A.7.39}$$

$$\delta S_1=2\pi R^2\sin\alpha\mathrm{d}\alpha, \tag{A.7.40}$$

which implies that

$$\delta S_0=\delta S_1\cos\alpha. \tag{A.7.41}$$

Substituting Eq. (A.7.41) into Eq. (A.7.38) wyields

$$l_1=l_0\cos^{1/2}\alpha. \tag{A.7.42}$$

Note that $\cos\alpha=k_z/k$. A comparison with Eq. (A.7.29) and Eq. (A.7.13) reveals that we must take an additional factor of R into account [99], resulting in

$$\vec{a}\left(k_x,k_y\right)=R\left(k_z/k\right)^{1/2}\mathbf{M}\,\vec{E}_{\mathrm{i}}\left(k_x,k_y\right). \tag{A.7.43}$$

Finally, we can rewrite Eq. (A.7.30) in the form

$$\vec{E}(\vec{r}) = -\frac{\mathrm{i}R}{2\pi}\iint_{\Omega_1}\frac{\mathbf{M}\vec{E}_\mathrm{i}\left(k_x,k_y\right)}{\sqrt{k_z k}}\mathrm{e}^{\mathrm{i}\vec{k}\cdot\vec{r}}\mathrm{d}k_x\mathrm{d}k_y. \qquad \text{(A.7.44)}$$

With respect to the numerical implementation, it is convenient to interpret the integration over $\mathrm{d}k_x$ and $\mathrm{d}k_y$ as a two-dimensional inverse Fourier transform and use the corresponding FFT algorithm for the computation of the integral [78]:

$$\vec{E}(\vec{r}) \propto \mathrm{FFT}^{-1}\left[\frac{\mathbf{M}\vec{E}_\mathrm{i}\left(k_x,k_y\right)}{\sqrt{k_z}}\mathrm{e}^{\mathrm{i}k_z z}\right]. \qquad \text{(A.7.45)}$$

A single FFT is performed on an entire *xy*-plane at a specific *z*-coordinate. As a result, the FFT is more than 100x faster than a point-wise direct numerical integration [100].

If the image space consists of a single homogeneous medium, the lens operator $\mathbf{M}$ is a simple 3 x 3 matrix which rotates the electric field vectors. It is straightforward to extend this approach to stratified media [76]. In this case, a set of directionally dependent propagation matrices accounts for reflections at the interface transitions.

– [End of quotation] –

References

[1] "Karlsruhe Institute of Technology", "Phoibos Project Homepage." [Online]. Available: www.phoibos.de. [Accessed: 15-May-2016].

[2] Mellanox Technologies, "Silicon based variable optical attenuator." [Online]. Available:

http://www.mellanox.com/page/optical_interconnects_voa?mtag=interconnect_voa. [Accessed: 03-Feb-2015].

[3] D. Taillaert, W. Bogaerts, P. Bienstman, T. F. Krauss, P. Van Daele, I. Moerman, S. Verstuyft, K. De Mesel, and R. Baets, "An out-of-plane grating coupler for efficient butt-coupling between compact planar waveguides and single-mode fibers," *IEEE J. Sel. Top. Quantum Electron.*, vol. 38, no. 7, pp. 949–955, 2002.

[4] F. Van Laere, G. Roelkens, M. Ayre, J. Schrauwen, D. Taillaert, D. Van Thourhout, T. F. Krauss, and R. Baets, "Compact and highly efficient grating couplers between optical fiber and nanophotonic waveguides," *IEEE J. Light. Technol.*, vol. 25, no. 1, pp. 151–156, 2007.

[5] D. Vermeulen, S. Selvaraja, P. Verheyen, G. Lepage, W. Bogaerts, P. Absil, D. Van Thourhout, and G. Roelkens, "High-efficiency fiber-to-chip grating couplers realized using an advanced CMOS-compatible Silicon-On-Insulator platform," *Opt. Express*, vol. 18, no. 17, pp. 18278–18283, Aug. 2010.

[6] X. Chen, C. Li, C. K. Y. Fung, S. M. G. Lo, and H. K. Tsang, "Apodized waveguide grating couplers for efficient coupling to optical fibers," *IEEE Photonics Technol. Lett.*, vol. 22, no. 15, pp. 1156–1158, 2010.

[7] D. Taillaert, H. Chong, P. I. Borel, L. H. Frandsen, R. M. D. La Rue, and R. Baets, "A compact two-dimensional grating coupler used as a polarization splitter," *IEEE Photonics Technol. Lett.*, vol. 15, no. 9, pp. 1249–1251, 2003.

[8] C. Gunn, "CMOS photonics for high-speed interconnects," *IEEE Micro*, vol. 26, no. 2, pp. 58–66, 2006.

[9] Y. Tang, Z. Wang, L. Wosinski, U. Westergren, and S. He, “Highly efficient nonuniform grating coupler for silicon-on-insulator nano-photonic circuits,” *Opt. Lett.*, vol. 35, no. 8, pp. 1290–1292, Apr. 2010.

[10] S. Scheerlinck, J. Schrauwen, F. Van Laere, D. Taillaert, D. Van Thourhout, and R. Baets, “Efficient, broadband and compact metal grating couplers for silicon-on-insulator waveguides.,” *Opt. Express*, vol. 15, no. 15, pp. 9625–9630, 2007.

[11] D. Taillaert, F. Van Laere, M. Ayre, W. Bogaerts, D. Van Thourhout, P. Bienstman, and R. Baets, “Grating couplers for coupling between optical fibers and nanophotonic waveguides,” *Jpn. J. Appl. Phys.*, vol. 45, no. 8A, pp. 6071–6077, 2006.

[12] W. S. Zaoui, A. Kunze, W. Vogel, M. Berroth, J. Butschke, F. Letzkus, and J. Burghartz, “Bridging the gap between optical fibers and
silicon photonic integrated circuits,” *Opt. Express*, vol. 22, no. 2, pp. 1277–1286, Jan. 2014.

[13] V. R. Almeida, R. R. Panepucci, and M. Lipson, “Nanotaper for compact mode conversion,” *Opt. Lett.*, vol. 28, no. 15, pp. 1302–1304, Aug. 2003.

[14] R. Dangel, J. Hofrichter, F. Horst, D. Jubin, A. La Porta, N. Meier, I. M. Soganci, J. Weiss, and B. J. Offrein, “Polymer waveguides for electro-optical integration in data centers and high-performance computers,” *Opt. Express*, vol. 23, no. 4, pp. 4736–4750, Feb. 2015.

[15] T. Shoji, T. Tsuchizawa, T. Watanabe, K. Yamada, and H. Morita, “Low loss mode size converter from 0.3 um square Si wire waveguides to singlemode fibres,” *Electron. Lett.*, vol. 38, no. 25, pp. 1669–1670, 2002.

[16] S. McNab, N. Moll, and Y. Vlasov, “Ultra-low loss photonic integrated circuit with membrane-type photonic crystal waveguides,” *Opt. Express*, vol. 11, no. 22, pp. 2927–2939, Nov. 2003.

[17] L. C. L. Chen, C. R. Doerr, Y.-K. C. Y.-K. Chen, and T.-Y. L. T.-Y. Liow, “Low-loss and broadband cantilever couplers between standard cleaved fibers and high-index-contrast SiN or Si waveguides,” *IEEE Photonics Technol. Lett.*, vol. 22, no. 23, pp. 1744–1746, 2010.

[18] F. E. Doany, B. G. Lee, S. Assefa, W. M. J. Green, M. Yang, C. L. Schow, C. V Jahnes, S. Zhang, J. Singer, V. I. Kopp, J. A. Kash, and Y. A. Vlasov, "Multichannel high-bandwidth coupling of ultradense silicon photonic waveguide array to standard-pitch fiber array," *J. Light. Technol.*, vol. 29, no. 4, pp. 475–482, Feb. 2011.

[19] R. Soref, "The past, present, and future of silicon photonics," *IEEE J. Sel. Top. Quantum Electron.*, vol. 12, no. 6, pp. 1678–1687, 2006.

[20] M. Hochberg and T. Baehr-Jones, "Towards fabless silicon photonics," *Nat Phot.*, vol. 4, no. 8, pp. 492–494, Aug. 2010.

[21] R. Nagarajan, M. Kato, J. Pleumeekers, P. Evans, S. Corzine, S. Hurtt, A. Dentai, S. Murthy, M. Missey, R. Muthiah, R. A. Salvatore, C. Joyner, R. Schneider, M. Ziari, F. Kish, and D. Welch, "InP photonic integrated circuits," *Selected Topics in Quantum Electronics, IEEE Journal of*, vol. 16, no. 5. pp. 1113–1125, 2010.

[22] P. C. Schindler, D. Korn, C. Stamatiadis, M. F. O'Keefe, L. Stampoulidis, R. Schmogrow, P. Zakynthinos, R. Palmer, N. Cameron, Y. Zhou, R. G. Walker, E. Kehayas, S. Ben-Ezra, I. Tomkos, L. Zimmermann, K. Petermann, W. Freude, C. Koos, and J. Leuthold, "Monolithic GaAs electro-optic IQ modulator demonstrated at 150 Gbit/s with 64QAM," *J. Light. Technol.*, vol. 32, no. 4, pp. 760–765, 2014.

[23] A. Leinse, R. G. Heideman, E. J. Klein, R. Dekker, C. G. H. Roeloffzen, and D. A. I. Marpaung, "TriPleX: platform technology for photonic integration: Applications from UV through NIR to IR," *Information Photonics (IP), 2011 ICO International Conference on*. pp. 1–2, 2011.

[24] H. Takahashi, "Recent progress on silica-based planar lightwave circuit for flexible and transparent photonic networks," *Photonics in Switching, 2008. PS 2008. International Conference on*. pp. 1–2, 2008.

[25] S. Maruo, O. Nakamura, and S. Kawata, "Three-dimensional microfabrication with two-photon-absorbed photopolymerization," *Opt. Lett.*, vol. 22, no. 2, p. 132, Jan. 1997.

[26] C. N. LaFratta, J. T. Fourkas, T. Baldacchini, and R. A. Farrer, "Multiphoton fabrication," *Angewandte Chemie - International Edition*, vol. 46, no. 33. pp. 6238–6258, 2007.

[27] G. von Freymann, A. Ledermann, M. Thiel, I. Staude, S. Essig, K. Busch, and M. Wegener, "Three-dimensional nanostructures for photonics," *Adv. Funct. Mater.*, vol. 20, no. 7, pp. 1038–1052, 2010.

[28] G. Odian, *Principle of Polymerization*, 4th ed. New York, 2004.

[29] Y. R. Shen, "The principles of nonlinear optics," *Opt. Laser Technol.*, vol. 17, no. 2, p. 104, Apr. 1984.

[30] M. Deubel, G. Von Freymann, M. Wegener, S. Pereira, K. Busch, and C. M. Soukoulis, "Direct laser writing of three-dimensional photonic-crystal templates for telecommunications," *Nat Mater*, vol. 3, no. 7, pp. 444–447, Jul. 2004.

[31] B. Richter, T. Pauloehrl, J. Kaschke, D. Fichtner, J. Fischer, A. M. Greiner, D. Wedlich, M. Wegener, G. Delaittre, C. Barner-Kowollik, and M. Bastmeyer, "Three-dimensional microscaffolds exhibiting spatially resolved surface chemistry," *Adv. Mater.*, vol. 25, no. 42, pp. 6117–6122, Nov. 2013.

[32] J. Bauer, S. Hengsbach, I. Tesari, R. Schwaiger, and O. Kraft, "High-strength cellular ceramic composites with 3D microarchitecture," *Proc. Natl. Acad. Sci.*, vol. 111, no. 7, pp. 2453–2458, Feb. 2014.

[33] G. Schmid, W. R. Leeb, G. Langer, V. Schmidt, and R. Houbertz, "Gbit/s transmission via two-photon-absorption-inscribed optical waveguides on printed circuit boards," *Electron. Lett.*, vol. 45, no. 4, pp. 219–221, 2009.

[34] R. Houbertz, V. Satzinger, V. Schmid, W. Leeb, and G. Langer, "Optoelectronic printed circuit board: 3D structures written by two-photon absorption," in *Organic 3D Photonics Materials and Devices II*, 2008, vol. 7053, no. 1, p. 70530B.

[35] S. Uhlig, L. Frohlich, M. Chen, N. Arndt-Staufenbiel, G. Lang, H. Schröder, R. Houbertz, M. Popall, M. Robertsson, and L. Fröhlich, "Polymer optical interconnects: a scalable large-area panel processing approach," *IEEE Trans. Adv. Packag.*, vol. 29, no. 1, pp. 158–170, Feb. 2006.

[36] R. Houbertz, H. Wolter, P. Dannberg, J. Serbin, and S. Uhlig, "Advanced packaging materials for optical applications: bridging the gap between nm-size structures and large-area panel processing," 2006, vol. 6126, pp. 612605–612613.

[37] F. ISC, “Ormocere.” [Online]. Available: http://www.ormocere.de. [Accessed: 22-Dec-2016].

[38] MicroChem Corp., “MicroChem Company Home Page.” [Online]. Available: http://www.microchem.com/Prod-SU8_KMPR.htm.

[Accessed: 28-Dec-2015].

[39] Nanoscribe GmbH, “Nanoscribe company homepage.” [Online]. Available: www.nanoscribe.de. [Accessed: 28-Dec-2015].

[40] S. S. M. Kuebler, M. Rumi, T. Watanabe, K. Braun, B. H. Cumpston, A. A. Heikal, L. L. Erskine, S. Thayumanavan, S. Barlow, S. R. Marder, and J. W. Perry, “Optimizing two-photon initiators and exposure conditions for three-dimensional lithographic microfabrication,” *J. Photopolym. Sci. Technol.*, vol. 14, no. 4, pp. 657–668, 2001.

[41] M. Thiel and H. Fischer, “Method and device for a spatially resolved introduction of an intensity pattern comprising electro-magnetic radiation into a photosensitive substance as well as applications thereof,” US20120218535 A1, 2012.

[42] W. K. Pratt, *Digital image processing: PIKS inside*, vol. 5. John Wiley & Sons, Inc., 2001.

[43] W. Burger and M. J. Burge, *Digital image processing: An algorithmic introduction using Java*, 1st ed. Springer, 2008.

[44] M. Heiblum and J. Harris, “Analysis of curved optical waveguides by conformal transformation,” *IEEE J. Quantum Electron.*, vol. 11, no. 2, 1975.

[45] K. R. Hiremath, M. Hammer, R. Stoffer, L. Prkna, and J. Čtyroký, “Analytic approach to dielectric optical bent slab waveguides,” in *Optical and Quantum Electronics*, 2005, vol. 37, no. 1–3, pp. 37–61.

[46] E. A. J. Marcatili, “Bends in optical dielectric guides,” *Bell Syst. Tech. J.*, vol. 48, no. 7, pp. 2103–2132, 1969.

[47] E. A. J. Marcatili and S. E. Miller, “Improved relations describing directional control in electromagnetic wave guidance,” *Bell Syst. Tech. J.*, vol. 48, no. 7, pp. 2161–2188, 1969.

[48] G. Grau and W. Freude, *Optische Nachrichtentechnik: eine Einführung*. Berlin: Springer, 1991.

[49] R. G. Hunsperger, *Integrated optics: theory and technology*. Springer, 2009.

[50] A. Borreman, S. Musa, A. A. M. Kok, M. B. J. Diemeer, and A. Driessen, "Fabrication of polymeric multimode waveguides and devices in SU-8 photoresist using selective polymerization.," in *IEEE/LEOS Benelux Chapter 2002 Annual Symposium*, 2002, pp. 83–86.

[51] C. Koos, *Nanophotonic devices for linear and nonlinear optical signal processing*. Karlsruhe: Universitätsverlag Karlsruhe, 2007.

[52] L. Papula, Ed., *Mathematik für Ingenieure und Naturwissenschaftler : Band 3 Vektoranalysis, Wahrscheinlichkeitsrechnung, Mathematische Statistik, Fehler- und Ausgleichsrechnung*, 5. Wiesbaden: Vieweg+Teubner Verlag, 2008.

[53] R. Fletcher, "A new approach to variable metric algorithms," *Comput. J.*, vol. 13, no. 3, pp. 317–322, Mar. 1970.

[54] the MathWorks Inc., "Description of the function 'fminunc.'" [Online]. Available:

http://de.mathworks.com/help/optim/ug/fminunc.html.

[Accessed: 29-Jan-2015].

[55] the MathWorks Inc., "Description of the 'BFGS' method." [Online]. Available: http://de.mathworks.com/help/optim/ug/unconstrained-nonlinear-optimization-algorithms.html#brnpcye. [Accessed: 03-Nov-2015].

[56] Nanoscribe GmbH, "DeScribe." [Online]. Available: http://www.nanoscribe.de/en/products/software/. [Accessed: 17-Oct-2016].

[57] D. Miller, "Device requirements for optical interconnects to silicon chips," *Proc. IEEE*, vol. 97, no. 7, pp. 1166–1185, 2009.

[58] ITRS.net, "Interconnects Roadmap," *2009*. [Online]. Available: http://www.itrs.net/Links/2009ITRS/2009Chapters_2009Tables/2009_Interconnect.pdf. [Accessed: 01-Jan-2010].

[59] AGC Chemicals Europe, "AGC Chemicals company hompage." [Online]. Available: http://www.agcce.com/Cytop.asp. [Accessed: 28-Dec-2015].

[60] CST-Comupter Simulation Technology, "News in EM thermal co-simulation," 2007. [Online]. Available:

https://www.cst.com/Content/Events/UGM2009/5-3-1-News-in-EM-Thermal-Co-Simulation.pdf. [Accessed: 12-Dec-2015].

[61] CST - Comupter Simulation Technology, "CST company homepage." [Online]. Available: https://www.cst.com/Products/CSTMWS. [Accessed: 28-Dec-2015].

[62] D. Feng, W. Qian, H. Liang, N.-N. Feng, S. Liao, C.-C. Kung, J. Fong, Y. Liu, R. Shafiiha, D. C. Lee, B. J. Luff, and M. Asghari, "Terabit/s single chip WDM receiver on the SOI platform," in *Group IV Photonics (GFP), 2011 8th IEEE International Conference on*, 2011, pp. 320–322.

[63] R. Schmogrow, R. Bouziane, M. Meyer, P. A. Milder, P. C. Schindler, P. Bayvel, R. Killey, W. Freude, and J. Leuthold, "Real-time digital nyquist-WDM and OFDM signal generation: spectral efficiency versus DSP complexity," in *European Conference and Exhibition on Optical Communication*, 2012, p. Mo.2.A.4.

[64] D. Hillerkuss, R. Schmogrow, T. Schellinger, M. Jordan, M. Winter, G. Huber, T. Vallaitis, R. Bonk, P. Kleinow, F. Frey, M. Roeger, S. Koenig, A. Ludwig, A. Marculescu, J. Li, M. Hoh, M. Dreschmann, J. Meyer, S. Ben Ezra, N. Narkiss, B. Nebendahl, F. Parmigiani, P. Petropoulos, B. Resan, A. Oehler, K. Weingarten, T. Ellermeyer, J. Lutz, M. Moeller, M. Huebner, J. Becker, C. Koos, W. Freude, and J. Leuthold, "26 Tbit s−1 line-rate super-channel transmission utilizing all-optical fast Fourier transform processing," *Nat. Photonics*, vol. 5, no. 6, pp. 364–371, Jun. 2011.

[65] R. Houbertz, G. Domann, C. Cronauer, A. Schmitt, H. Martin, J.-U. Park, L. Fröhlich, R. Buestrich, M. Popall, U. Streppel, P. Dannberg, C. Wächter, and A. Bräuer, "Inorganic–organic hybrid materials for application in optical devices," *Thin Solid Films*, vol. 442, no. 1–2, pp. 194–200, Oct. 2003.

[66] S. Fessel, A. M. Schneider, S. Steenhusen, R. Houbertz, and P. Behrens, "Towards an atomistic model for ORMOCER®-I: Application of forcefield methods," *J. Sol-Gel Sci. Technol.*, vol. 63, no. 3, pp. 356–365, 2012.

[67] S. Steenhusen, T. Stichel, R. Houbertz, and G. Sextl, "Multi-photon polymerization of inorganic-organic hybrid polymers using visible or IR ultrafast laser pulses for optical or optoelectronic devices," 2010, vol. 7591, pp. 759112–759114.

[68] S. Feng, Y. Geng, K. M. Lau, and A. W. Poon, "Epitaxial III-V-on-silicon waveguide butt-coupled photodetectors," in *IEEE International Conference on Group IV Photonics GFP*, 2012, pp. 51–53.

[69] H. Park, M. N. Sysak, H. W. Chen, A. W. Fang, D. Liang, L. Liao, B. R. Koch, J. Bovington, Y. Tang, K. Wong, M. Jacob-Mitos, R. Jones, and J. E. Bowers, "Device and integration technology for silicon photonic transmitters," *IEEE J. Sel. Top. Quantum Electron.*, vol. 17, no. 3, pp. 671–688, 2011.

[70] R. E. Camacho-Aguilera, Y. Cai, N. Patel, J. T. Bessette, M. Romagnoli, L. C. Kimerling, and J. Michel, "An electrically pumped germanium laser.," *Opt. Express*, vol. 20, no. 10, pp. 11316–20, May 2012.

[71] S. Tanaka, T. Matsumoto, T. Kurahashi, M. Matsuda, A. Uetake, S. Sekiguchi, Y. Tanaka, and K. Morito, "Flip-chip-bonded, 8-wavelength AlGaInAs DFB laser array operable up to 70°C for silicon WDM interconnects," in *European Conference on Optical Communication*, 2014, pp. 1–3.

[72] K. Okamoto, *Fundamentals of Optical Waveguides*. San Diego: Elsevier, 2006.

[73] Compound Semiconductor Technologies Global Ltd, "Compound Semiconductor Technologies company homepage," 2015. [Online]. Available: http://www.compoundsemi.co.uk/. [Accessed: 28-Dec-2015].

[74] M. L. Goedecke, "Coupling of photonic wire bonds to vertical device facets," Karlsruhe Institute of Technology, 2015.

[75] J. J. Stamnes, *Waves in focal regions: propagation, diffraction and fpcusing of light, sound and water waves (optics & optoelectronics)*. Institute of Physics Pub., 1986.

[76] A. S. Van De Nes, L. Billy, S. F. Pereira, J. J. M. Braat, A. S. van de Nes, L. Billy, S. F. Pereira, and J. J. M. Braat, "Calculation of the vectorial field distribution in a stratified focal region of a high numerical aperture imaging system," *Opt. Express*, vol. 12, no. 7, p. 1281, Apr. 2004.

[77] A. S. van de Nes, "Rigorous electromagnetic field calculations for advanced optical systems," Delft University of Technology, 2005.

[78] J. Fischer, "Three-dimensional optical lithography beyond the diffraction limit," Karlsruhe Institute of Technology, 2012.

[79] A. F. Oskooi, D. Roundy, M. Ibanescu, P. Bermel, J. D. Joannopoulos, and S. G. Johnson, "Meep: A flexible free-software package for electromagnetic simulations by the FDTD method," *Comput. Phys. Commun.*, vol. 181, no. 3, pp. 687–702, Mar. 2010.

[80] Epoxy Technology Inc, "Epoxy Technology Inc company homepage." [Online]. Available: http://www.epotek.com/site/german.html. [Accessed: 08-Dec-2015].

[81] Corning Incorporated, "Corning ® SMF-28e +® Optical Fiber," *Datasheet*. [Online]. Available:

https://www.corning.com/media/worldwide/csm/documents/Corning SMF28e+® Photonic Specialty Fiber.pdf. [Accessed: 28-Dec-2015].

[82] I. H. Malitson, "Interspecimen comparison of the refractive index of fused silica," *J. Opt. Soc. Am.*, vol. 55, no. 10, p. 1205, Oct. 1965.

[83] Dymax, "Dymax company homepage." [Online]. Available: http://www.dymax.com/index.php/de/home-de. [Accessed: 28-Dec-2015].

[84] Y.-T. Tseng, T.-Y. Hung, and J.-H. Liu, "A fiber splicing-plane-inspection technique using lens–fiber interference for the cascaded fiber fabrication," *IEEE J. Light. Technol.*, vol. 25, no. 3, pp. 803–810, Mar. 2007.

[85] H. Takara, A. Sano, T. Kobayashi, H. Kubota, H. Kawakami, A. Matsuura, Y. Miyamoto, Y. Abe, H. Ono, K. Shikama, Y. Goto, K. Tsujikawa, Y. Sasaki, I. Ishida, K. Takenaga, S. Matsuo, K. Saitoh, M. Koshiba, and T. Morioka, "1.01-Pb/s (12 SDM/222 WDM/456 Gb/s) crosstalk-managed zransmission with 91.4-b/s/Hz aggregate spectral efficiency," in *European Conference and Exhibition on Optical Communication*, 2012, p. Th.3.C.1.

[86] C. R. Doerr and T. F. Taunay, "Silicon photonics core-, wavelength-, and polarization-diversity receiver," *IEEE Photonics Technol. Lett.*, vol. 23, no. 9, pp. 597–599, 2011.

[87] D. Marcuse, "Bending losses of the asymmetric slab waveguide," *Bell Syst. Tech. J.*, vol. 50, no. 8, pp. 2551–2563, Oct. 1971.

[88] A. D. Yablon, "Multifocus tomographic algorithm for measuring optically thick specimens," *Opt. Lett.*, vol. 38, no. 21, pp. 4393–6, Nov. 2013.

[89] H. Sun and S. Kawata, "Two-photon photopolymerization and 3D lithographic microfabrication," *Quantum*, vol. 170, pp. 169–274, 2004.

[90] Microchemicals GmbH, "Microchemicals product homepage." [Online]. Available:

http://www.microchemicals.com/de/produkte/haftvermittler/ti_prime.h tml. [Accessed: 28-Dec-2015].

[91] J. D. Jackson, *Classical Electrodynamics*, 3rd ed. New York: Wiley, 1998.

[92] A. E. Siegman, *Lasers*. Sausalito, California: University Science Books, 1986.

[93] W. Freude and G. Grau, "Rayleigh-Sommerfeld and Helmholtz-Kirchhoff integrals: application to the scalar and vectorial theory of wave propagation and diffraction," *IEEE J. Light. Technol.*, vol. 13, no. 1, pp. 24–32, 1995.

[94] E. Wolf, "Electromagnetic Diffraction in Optical Systems. I. An Integral Representation of the Image Field," *Proc. R. Soc. A Math. Phys. Eng. Sci.*, vol. 253, no. 1274, pp. 349–357, Dec. 1959.

[95] P. Debye, "Das Verhalten von Lichtwellen in der Nähe eines Brennpunktes oder einer Brennlinie," *Ann. Phys.*, vol. 335, no. 14, pp. 755–776, 1909.

[96] J. J. Stamnes, "Focusing of two-dimensional waves," *J. Opt. Soc. Am.*, vol. 71, no. 1, p. 15, 1981.

[97] Y. Li and E. Wolf, "Focal shifts in diffracted converging spherical waves," *Opt. Commun.*, vol. 39, no. 4, pp. 211–215, Oct. 1981.

[98] E. Wolf and Y. Li, "Conditions for the validity of the Debye integral representation of focused fields," *Opt. Commun.*, vol. 39, no. 4, pp. 205–210, Oct. 1981.

[99] B. Richards and E. Wolf, "Electromagnetic diffraction in optical systems II. Structure of the image field in an aplanatic system," *Proc. R. Soc. London Ser. a-Mathematical Phys. Sci.*, vol. 253, no. 1274, p. 3580379, Dec. 1959.

[100] M. Leutenegger, R. Rao, R. A. Leitgeb, and T. Lasser, "Fast focus field calculations," *Opt. Express*, vol. 14, no. 23, pp. 4897–4903, 2006.

[101] S. Maruo and K. Ikuta, "Three-dimensional microfabrication by use of single-photon-absorbed polymerization," *Appl. Phys. Lett.*, vol. 76, no. 91, pp. 2656–2658, 2000.

[102] J. B. Mueller, J. Fischer, F. Mayer, M. Kadic, and M. Wegener, "Polymerization kinetics in three-dimensional direct laser writing," *Adv. Mater.*, vol. 26, no. 38, pp. 6566–6571, 2014.

[103] P. Mueller, M. Thiel, and M. Wegener, "3D direct laser writing using a 405 nm diode laser," *Opt. Lett.*, vol. 39, no. 24, pp. 6847–6850, 2014.

[104] M. Thiel, J. Fischer, G. von Freymann, and M. Wegener, "Direct laser writing of three-dimensional submicron structures using a continuous-wave laser at 532 nm," *Appl. Phys. Lett.*, vol. 97, no. 22, p. 221102, 2010.

[105] L. Li and J. T. Fourkas, "Multiphoton Polymerization," *Mater. Today*, vol. 10, no. 6, pp. 30–37, Jun. 2007.

List of Symbols

Greek Symbols

α	power attenuation coefficient
α	half opening angle of the focused beam
$\chi^{(3)}$	nonlinear susceptibility of third order, complex scalar
Δ	refractive index difference
Δx	line distance along the x-axis
Δy	line distance along the y-axis
Δz	line distance along the z-axis
Δt	simulation time step
ε_0	dielectric constant
δS	surface element
η	overlap integral
γ	fitting parameter
κ	curvature
λ	wavelength
μ	second central moment
$\phi_0(x,\mathrm{y})$	optical aberration function
σ	standard deviation of the cumulative distribution function
θ	angle between camera and lithography coordinates
τ	pulse duration
ω	optical frequency
Ω_1	solid angle
∇	$\left(\frac{\partial}{\partial x},\frac{\partial}{\partial y},\frac{\partial}{\partial z}\right)$ Nabla Operatror

Latin Symbols

$\mathcal{A}$	aperture area
$\mathbf{A}$	matrix containing the gray scale values of every camera pixel
$\vec{a}(k_x,k_y)$	vectorial amplitude of the electric field
a,b	pixel dimesions in [µm]
$\vec{b}(k_x,k_y)$	vectorial amplitude of the magnetic field
$\vec{B}(\vec{r},t)$	magnetic flux density
c	speed of light
$c_{\mathrm{conversion}}$	loss at interface between straight and bent waveguide
$\vec{D}(\vec{r},t)$	dielectric displacement
d_{PWB}	diameter
d	distance between geometrical focus and device facet
$\vec{\mathcal{E}}(\vec{r},t)$	electrical field vector
$\vec{E}(\vec{r})$	time independent electric field
$\vec{E}_{\mathrm{S}}$	electrical field in the stationary point
$(\vec{e}_x,\vec{e}_y,\vec{e}_{\mathrm{z}})$	Cartesian base vectors
$(\vec{e}_{\mathrm{r}},\vec{e}_{\phi},\vec{e}_{\mathrm{z}})$	zylindrical basis vectors
$\mathcal{F}_{k_x,k_y}\{\ \}$	Fourier transformation
f_{rep}	repetition rate
$g(x,y,z)$	penalty function
$g(z)$	overall edge strength *vs.* focus position
$\mathbf{G}_{\mathrm{x}},\mathbf{G}_{\mathrm{y}}$	horizontal/vertical gradient matrix
$\vec{\mathcal{H}}(\vec{r},t)$	magnetic field
$\vec{H}(\vec{r})$	time independent contribution of the magnetic field
h	waveguide height
i	imaginary unit
IDx, IDy	pixel values in x,y direction

$\mathrm{Im}\{x\}$	imaginary part of a complex number x
$J(\vec{r})$	loss functional
$\vec{k}$	propagation vector
l	length
l_0^2	energy density
L	number of equidistant nodes
M	number of layers used for shape conversion
$\mathbf{M}$	transformation matrix
$\mathbf{M}_{\mathrm{LC}}$	rotation matrix
N	number of contour points in a layer
$\mathbf{N}$	rotation matrix
n	refractive index
n_{min}	minimum refractive index in simulation domain
n_{co}	core refractive index
n_{cl}	cladding refractive index
P	optical power
$\mathbf{P}$	coordinate transformation matrix
$\vec{P}$	polarization
$P_1 \ldots P_4$	predefined positions in the lithography or piezo x-y-plane
p	two-photon absorption rate
q	fitting parameter
$\mathrm{Re}\{x\}$	real part of a complex number x
R, R_{c}	radius
$\vec{r}$	position vector
$\vec{r}(t)$	trajectory
$\vec{r}_c$	position vector in camera coordintes
$\vec{r}_{\mathrm{L}}$	position vector in lithography coordinates

$\vec{r}_{LC}$	translation between two images in image stitching
$\vec{r}_{\mathrm{C},0}$	offset between camera and lithography coordinates
r,φ,z	Cylindrical coordinate axis
$\mathbf{S}_{\mathrm{x}},\mathbf{S}_{\mathrm{y}}$	Sobel gradient operator
s	arc length
S	Courant factor
$\mathbf{T}_{\mathrm{LC}}$	transformation between camera and lithography coordinates
$\mathbf{T}_{\mathrm{CL}}$	backward transformation
t	curve parameter
$\vec{U}_0$	voltage
v	writing velocity
$\vec{v}(t)$	tangential vector of trajectory
$\vec{v}_0$	tangential vector of trajectory in the starting position
$\vec{v}_1$	tangential vector of trajectory in ending position
w	waveguide width
W_{el}	electric field energy
x,y,z	Cartesian axis
z_{L}	height of foculs plane of the lithography coordinate system
z_{LC}	offset between camera and laser focus plane

Acronyms

AOM	Accusto-optical-modulator
AWG	Arrayed waveguide grating
BER	Bit error probability
BFGS	Broyden-Fletcher-Goldfarb-Shannon algorithm
BPM	Beam propagation method
BS	Beam splitter
BTB	Back-to-back, i.e. without device
CMOS	Complementary metal oxide semiconductor
DBS	Dichroic beam splitter
DFB	Distributed-feedback laser
DIL	Disinterleaver
DUT	Device under test
EVM	Error vector magnitude
FEC	Forward error compensation
FIT	Finite integration technique
FFT	Fast Fourier transformation
GC	Grating coupler
HCSEL	Horizontal-cavity surface-emitting laser
HNLF	Highly nonlinear fiber
IC	Integrated circuit
LED	Light-emitting diode
MCF	Multi-core fiber
MLL	Mode-locked laser
NA	Numerical aperture
Nyquist WDM	Nyquist wavelength division multiplexing
OMA	Optical modulation analyzer

OOK	On-off-keying
OSA	Optical spectrum analyzer
PIC	Photonic integrated circuits
PMD	Polarization mode dispersion
PRBS	Pseudo-random bit sequence
PWB	Photonic Wire Bond
RZ	Return-to-zero
SEM	Scanning electron microscope
SDM	Space-division multiplexing
SOI	Silicon-on-Insulator
SMF	Standard Single mode fiber
stl	Standard Tessellation Language
TE	Transverse electric
TM	Transverse magnetic
TPP	Two-Photon Polymerization
VCSEL	Vertical cavity surface emitting laser
WDM	Wavelength division multiplexing
WG	Waveguide
WSS	Wavelength selective switch
3D	Three-dimensional
3D DWL	Three-dimensional direct laser writing
16QAM	16ary quadrature amplitude modulation
64QAM	64ary quadrature amplitude modulation

Acknowledgements

Die hier vorliegende Dissertationsschrift entstand basierend auf meiner Arbeit am Institut für Photonik und Quantenelektronik (IPQ) und am Institut für Mikrostrukturtechnik (IMT) des Karlsruher Instituts für Technologie (KIT). Die Arbeit war eingebunden in das vom Bundesministerium für Bildung und Forschung geförderten Projektes PHOIBOS unter dem Förderkennzeichen AZK 13N12574, des Verbundprojekt "Polymere Lichtwellenleiter für die Nanophotonik" (POLINA) das durch das Bundesministerium für Wirtschaft und Technologie unter der Fördernummer BMWi-AIF (ZIM) KF 2308201MF9 gefördert wurde, sowie in das Projekt A4.07 des Zentrums für funktionale Nanostrukturen am Karlsruher Institut für Technologie.

Für die außerordentliche Zusammenarbeit und Unterstützung der letzten Jahre möchte ich mich bei allen denjenigen bedanken, die zum Gelingen der Arbeit beigetragen haben.

An erster Stelle danke ich meinem Doktorvater Prof. Dr. Christian Koos für das entgegengebrachte Vertrauen, die massive Unterstützung meiner Arbeit, das große und stetige Interesse an meiner Arbeit, die zahlreichen Anregungen und seine innovativen Ideen, und nicht zuletzt für die mir zugestandenen Freiheiten eigene Ideen zu verwirklichen.

Herrn Prof. Dr. Uli Lemmer danke ich für die Übernahme des Co-Referenten und stellvertretend für das ganze Lichttechnischen Institutes für die Unterstützung bei praktischen Dingen wie „Reinraum-Asyl“. Diese Unterstützung war für mich vor allem in der frühen Phase meiner Arbeit unersetzlich.

Bei Herrn Prof. Dr. Dr. Wolfgang Freude möchte ich mich ebenfalls für seine Rolle als Co-Referent bedanken. Sein großes Engagement, mit dem er meine Arbeit begleitet hat war eine wichtige Unterstützung für mich. Die Diskussionen mit ihm waren stets sehr hilfreich, lehrreich und zielführend.

Bei Herrn Prof. em. Dr. Gerhard Grau bedanke ich mich herzlich für seine Begleitung und moralische wie auch fachliche Unterstützung, sowie für das gute Vorbild, dass er für uns junge Wissenschaftler ist.

Meinen Projektpartnern Martin Hermattschweiler, Dr. Michael Thiel, Jörg Hoffmann, Philipp Simon, Philipp Hübner, Dr. Timo Mappes, Dr. Michael Totzeck, Dr. Stefan Richter, Dr. Bert Offrein, Dr. Ute Troppenz, Dr. Martin Möhrle, Dr. Werner Graudszus, Dr. Marcel Kroh, Lars Brettschneider und Lars Zimmermann danke ich für die schöne Zusammenarbeit im Phoibos Projekt.

Ute Troppenz, Dr. Karl Pfeiffer und Uwe Hollenbach ihnen sei ein Dank für die spannende Zusammenarbeit im Polina Projekt gesagt.

Herrn Prof. Dr. Weber, Herrn Prof. Dr. Schimmel und Herrn Prof. Dr. Wegener danke ich für die Unterstützung „meiner" Physik-Studenten.

Den Firmen Fibercore, CST Compound Semiconductors, Vertilas und Nanoscribe sage ich herzlich Danke für die wertvollen Diskussionen und das Bereitstellen einiger Bauteile und Verbrauchsmittel, ohne die es nicht gegangen wäre.

Ich danke Dr. Benjamin Richter, Sönke Steenhusen, Dr. Andrea Scheiwe und Dr. Jens Bauer für die gute „fachübergreifende" Zusammenarbeit und das großzügige Teilen euer Maschine, eures Lacks und eurer Erfahrungen.

Oswald Speck, Florian Rupp, Stefan Hengsbach, Marco Hummel, Martin Winkeler, David Guder, Manfred Hirsch, Werner Höhne, Manfred Bürger und den Azubis der feinmechanischen Werkstatt, ein herzliches Danke für die technische aber auch emotionale Unterstützung meiner Arbeit.

Bernadette Lehmann, Andrea Riemensperger, Tatjana Gassmann, Ursula Rothweiler, Angelika Olbrich, Zella Marchel, Anna-Maria Schubart und Ilse Kober danke ich für die administrative Unterstützung.

Meiner riesengroßen Studententruppe danke ich für die vielen Stunden harter aber auch schöner Arbeit, die wir zusammen verbracht haben. This special thank you goes to:

Inga Kaiser, mit der ich den ersten photonischen Wirebond überhaupt gebaut habe. Was für ein toller Moment als er tatschächlich funktioniert hat.

Julian Groß mit dem man leider nicht synchron schleußen konnte, dafür aber jede Menge tolle Experimente zu 2D Wellenleitern durchziehen konnte.

Simon Schneider, der von einenm super selbstständigen Studi zu einem prima Kollegen wurde. Danke, dass du mich beim OCT ein bisschen hast mitspielen lassen.

Gerhard Balthasar, mit dem ich zusammen die ersten Schritte in Richtung Automatisierung der Photonischen Wirebonds gegangen bin.

Kira Köhnle, mit der ich eine wünderbare Zeit während Studien- und Diplomarbeit verbringen durfte. Es leben die Plasmonen.

Denis Strauch, der den erstklassigen Spruch geprägt hat, dass „Technologieentwicklung eine Ausdauersportart ist“ was mittlerweile zu einer Art Arbeitsmotto für mich geworden ist.

Maher Rezem, der mit mir und Denis zusammen Polina zu einem schönen Abschluss gebracht hat und der mir viel über den Arabischen Frühling und seine Heimat Tunesien erklärt hat.

Meinen fleißigen und engagierten Hiwis Asiful Alam und Claus Gärthner die mich sowohl im Experiment als auch bei der Simulation toll unterstütz haben.

Tobias Hoose, der mit mir an einem Problem gearbeitet hat, für das andere Leute viel Geld bezahlt hätten und auf den ich mich so verlassen konnte wie auf wenige andere. Er teilt mein Kameltrauma und muss auch lachen wenn jemand „Johnny Walker“ sagt.

Jörg Hoffmann, der immer noch nicht viel redet, aber wenn, dann sollte man zuhören. Mit ihm habe ich an der Markererkennung und an der ersten Version der Faser-Chip-Kopplung gearbeitet und damit die Basis für einen reproduzierbaren Prozess geschaffen. Ich habe das große Glück ihn zum Kollegen zu haben und ihn einen Freund nennen zu dürfen.

Philipp Hübner, dem ich nicht nur das beste Brotrezept der Welt und die „Weltherrschaftspflanze“ verdanke, sondern auch jedem Menge Spaß aber auch Rückschläge bei der Ankopplung zwischen Laser und PWB.

Romain Bonjour, mit dem wir unser eigenes „der Gerät“ aufgebaut haben. Ein Projekt das zwar nicht in der erwarteten Weltherrschaft gemündet hat, bei dem wir aber unglaublich viel gelernt haben.

Stephan Dottermusch, mit dem ich an der PWB-Faserkopplung in planarer Anordnung weiter gearbeitet habe und mit dem ich gelernt habe, dass es durchaus auch Glasfasern gibt, die nicht für 1550 nm Licht geeignet sind.

Jonas Korckenberger, der mit seiner Simulation von gekrümmten Wellenleitern einiges an Einsichten geschaffen hat. To bend or not to bend that is the question.

Marcel Mauch, der unserem selbstgebauten Lithographieaufbau den absoluten Maßstab gegeben hat. Was hilft ein hochgenaues Gerät, wenn es nicht kalibriert ist?

Temitope Onanuga, der mit mir und Oding zusammen den Simulationsrechner ganz schön gequält hat um den Inteface-Verlusten an den Kragen zu gehen.

Maria-Laura Gödecke, mit der ich wohl von allen meinen Studis die meiste Zeit verbracht habe, nicht weil sie so viel Betreuung nötig gehabt hatte, sondern im Gegenteil, weil unsere Zusammenarbeit für mich eine sehr besondere war. Ihr mathematisches und physikalisches Verständnis gingen weit über das hinaus, was man selbst für einen sehr guten Studenten erwarten würde. Nur mit ihr zusammen war die Simulation und Implementierung der Kopplung von PWB an vertikale Bauteilfacetten überhaupt möglich. Unzählige Stunden waren wir zusammen im Mathe-Zimmer und im Reinraum, haben geflucht, gelacht und getüftelt. Ich kann kein Katzenvideo mehr anschauen, ohne dass ich an dich denke!

Manche Doktoranden haben das Glück einen besonderen Studenten betreuen zu dürfen, ich hatte dieses Glück SIEBZEHN MAL. Ihr ward der Hammer!

Meinen lieben (Ex-)Kollegen Tobias Hoose, Oding Billah, Dr. Robert Palmer, Dr. Philipp Schindler, Claudius Weimann, Simon Schneider, Aleksandar Nesic, Dr. Jörg Pfeifle, Matthias Lauermann, Dr. Argishti Melikyan, Sascha Mühlbrandt, Dr. Swen König, Dr. Sebastian Köber, Dr. Alexandra Ludwig, Djorn Karnick, Dr. Luca Alloatti, Dr. Dietmar Korn, Dr. René Schmogrow, Stefan Wolf, Dr. David Hillerkuss, Dr. Markus Winter, Dr. Arvind Mischra, Dr. Sean O'Duill, Dr. René Bonk, Dr. Thomas Vallaitis, Dr. Moritz Röger, Christos Klamouris, Frans Wegh, Kai Worms, Heinrich Sieber und Sebastian van der Ecken, danke ich für die prima Zusammenarbeit und die liebenswerte Arbeitsatmosphäre.

Für die Durchsicht meines Manuscripts bedanke ich mich herzlich bei Dr. Sebastian Köber, Dr. Thomas Vallaitis, Roman Reiner und Dr. Fabian Niesler. Ihr wart mir eine riesige Hilfe! Danke!

Vielen Dank an meine Freunde Eduard und Hannelore Weiler, euch möchte ich herzlich danken, durch euch weiß ich, dass Freundschaft keine Frage des Alters ist.

Vielen Dank an Monika und Peter Offenbacher (1952 - 2015), eure Tür war immer für mich offen, und euer Gastdozentenhaus war oft ein Zufluchtsort im Chaos für mich! Danke dafür! Peter, ich werde dich immer in Erinnerung behalten!

Meinen lieben Freunden Kerstin, Jan, Marina, Thorsten, und Nadine sei ein ganz spezieller Dank gesagt.

Meiner gesamten Familie, meinen lieben Omas Anneliese G. und Anneliese L. (1925-2013) und besonders meinem Schatz Tobi danke ich für die Unterstützung die ich stets erfahren habe und die auch in Durststrecken nie abgerissen ist. Gut Ding will eben Weile haben! Danke dafür, dass ihr das wisst.

Mama, Papa, Frank und Tobi, ihr seid die Besten.

List of Publications

Books

[B1] J. Leuthold, N. Lindenmann, C. Buchal, "35: Photonic Networks" in *Nanoelectronics and Information Technology,* 3rd Edition, Rainer Waser (Editor); John Wiley & Sons, Inc.; 2012; ISBN: 978-3-527-40927-3

Patents

[P1] T. Mappes, S. Koeber, N. Lindenmann, and C. Koos, "Herstellung von 3d-Freiform-wellenleiterstrukturen," WO Patent 2014 154 363 A1, March 28, 2014.

[P2] C. Koos, W. Freude, N. Lindenmann, and J. Leuthold, "Three-dimensional freeform waveguides for chip-chip connections," U.S. Patent 8 903 205 B2, Feburary 23, 2012.

[P3] C. Koos, W. Freude, N. Lindenmann, and J. Leuthold, "Method for producing photonic wire bonds," U.S. Patent 9 034 222 B2, February 23, 2012.

Journals

[J1] N. Lindenmann, S. Dottermusch, M. L. Goedecke, T. Hoose, M. R. Billah, T. P. Onanuga, A. Hofmann, W. Freude, and C. Koos, "Connecting silicon photonic circuits to multicore fibers by photonic wire bonding," *J. Light. Technol.*, vol. **33**, no. 4, pp. 755–760, 2015.

[J2] N. Lindenmann, G. Balthasar, D. Hillerkuss, R. Schmogrow, M. Jordan, J. Leuthold, W. Freude, C. Koos, "Photonic wire bonding a novel concept for chip-scale interconnects", *Opt. Express* **20**, 17667-17677 (2012).

[J3] D. Hillerkuss, R. Schmogrow, M. Meyer, S. Wolf, M. Jordan, P. Kleinow, N. Lindenmann, P. Schindler, A. Melikyan, X. Yang, S. Ben-Ezra, B. Nebendahl, M. Dreschmann, J. Meyer, F. Parmigiani, P. Petropoulos, B. Resan, A. Oehler, K. Weingarten, L. Altenhain, T. Ellermeyer, M. Moeller, M. Huebner, J. Becker, . Koos, W. Freude, J.

Leuthold "Single-laser 32.5 Tbit/s Nyquist WDM transmission" *IEEE J. Opt. Communications and Networking*, **4**, 715-723 (2012)

[J4] A. Melikyan, N. Lindenmann, S. Walheim, P. M. Leufke, S. Ulrich, J. Ye, P. Vincze, H. Hahn, Th. Schimmel, C. Koos, W. Freude, J. Leuthold, „Surface plasmon polarition absorption modulator", *Opt. Express* **19**, 8855-8869, (2011)

Conferences

[C1] T. Hoose, M. Billah, P. Marin, M. Blaicher, P. I. Dietrich, A. Hofmann, U. Troppenz, M. Möhrle, N. Lindenmann, M. Thiel, P. Simon, J. Hoffmann, M. L. Goedecke, W. Freude, and C. Koos, "Multi-chip integration by photonic wire bonding: Connecting surface and edge emitting lasers to silicon chips," in *Optical Fiber Communication Conference*, 2016, p. M2I.7.

[C2] P. Dietrich, S. Wondimu, T. Wienhold, M. Steidle, A. Hofmann, N. Lindenmann, M. R. Billah, T. Hoose, M. Blaicher, W. Freude, and C. Koos: "Herstellung mikrooptischer Komponenten durch Zwei-Photonen-Polymerisation," *in MikroSystemTechnik-Kongress „Von Bauelementen zu Systemen"*, 2015, paper EDAS Paper–ID 1570106799.

[C3] C. Koos, W. Freude, J. Leuthold, M. Kohl, L. R. Dalton, W. Bogaerts, M. Lauermann, S. Wolf, C. Weimann, A. Melikyan, N. Lindenmann, M. R. Billah, S. Muehlbrandt, S. Koeber, R. Palmer, K. Koehnle, L. Alloatti, D. L. Elder, A.-L. Giesecke, and T. Wahlbrink: "Silicon-organic hybrid (SOH) integration and photonic multi-chip systems: Extending the capabilities of the silicon photonic platform," *Photonics Conference (IPC)*, 2015. paper MF2.1, 2015.

[C4] M. Billah, T. Hoose, T. Onanuga, N. Lindenmann, P. Dietrich, T. Wingert, M. Goedecke, A. Hofmann, U. Troppenz, M. Moehrle, A. Sigmund, W. Freude, and C. Koos: "Multi-chip integration of lasers and silicon photonics by photonic wire bonding," in *CLEO: Science and Innovations*, CLEO-SI 2015, 2015, paper 2267.

[C5] C. Koos, W. Freude, R. Palmer, M. Lauermann, S. Koeber, J. Pfeifle, C. Weimann, P. C. Schindler, N. Lindenmann, T. Hoose, and M. R. Billah: "Silicon photonics, hybrid integration, and frequency combs:

technologies for Terabit/s communications, teratronics, and optical metrology," in *115th Annual DGaO Conference (DGaO'14),* 2014, paper H1.

[C6] N. Lindenmann, S. Dottermusch, M. L. Goedecke, T. Hoose, M. R. Billah, S. Koeber, W. Freude, C. Koos: "Connecting silicon photonic circuits to multi-core fibers by photonic wire bonding" in *Proceedings of IEEE Optical Interconnects Conference (OI'14)*, San Francisco (CA), USA, May 4-7, 2014, paper WD1

[C7] C. Koos, W. Freude, N. Lindenmann, S. Hoose, M. R. Billah: "Three-dimensional two-photon lithography: An enabling technology for photonic wire bonding and multi-chip integration" in *Proceedings of SPIE 8970 (LASE-SPIE'14), Laser 3D Manufacturing*, San Francisco (CA), March 6, 2014, USA, paper 897008 (Invited)

[C8] N. Lindenmann, T. Hoose, S. Steenhusen, M. R. Billah, S. Koeber, R. Houbertz, C. Koos: "Photonic wire bonding as an enabling technology for multi-chip photonic systems" in *Proceedings of SPIE OPTO (OPTO-SPIE'14)*, San Francisco (CA), USA, Feb. 1-6, 2014, paper 8991-5

[C9] C. Koos, W. Freude, J. Leuthold, L. Alloatti, R. Palmer, D. Korn, J. Pfeifle, P. Schindler, M. Lauermann, N. Lindenmann, S. Koeber, T. Hoose, M. R. Billah, H. Yu, W. Bogaerts, R. Baets, M. Fournier, J-M. Fedeli, R. Dinu, J. Bolten, T. Wahlbrink, M. Waldow,: "Silicon- organic hybrid (SOH) integration and photonic wire bonding: Enabling technologies for heterogeneous photonic systems", in *Proceedings of Optical Society of America 97th Annual Meeting (OSA'13)*, Orlando (FL), USA, October 05–10, 2013 Paper FTu1E.3 (Invited).

[C10] A. Nesic, L. Alloatti, J. Pfeifle, N. Lindenmann, J. Leuthold, W. Freude, C. Koos: "Highly efficient silicon-organic hybrid (SOH) phase shifters for reconfigurable ultra-fast signal processing" in *Proceedings of International Nano-Optoelectronics Workshop iNOW-2013*, Cargèse, France August 19 - August 30, 2013

[C11] C. Koos; W. Freeude, J. Leuthold, N. Lindenmann, S. Koeber, J. Hoffmann, T. Hoose, P. Huebner, M. Billah; „Photonic wire bonding: An enabling technology for heterogeneous multi-chip integration", *in Proceedings of OSA Integrated Photonics Research, Silicon and*

Nanophotonics (IPR'13), Rio Grande, Puerto Rico, USA, July 14–17, 2013, paper IM4A.3 **[invited]**

[C12] C. Koos, J. Leuthold, W. Freude, N. Lindenmann, S. Koeber, J. Hoffmann, T. Hoose, P. Huebner; „Photonic wire bonding: Nanophotonic interconnects fabricated by direct-write 3D lithography" *in Proceedings of 15th Intern. Conf. on Transparent Optical Networks* (ICTON'13), Cartagena, Spain, June 23–27, 2013 paper We.C2.4 **[invited]**

[C13] C. Koos, J. Leuthold, W. Freude, N. Lindenmann, S. Koeber, G. Balthasar, J. Hoffmann, T. Hoose, P. Huebner, D. Hillerkuss, R. Schmogrow; „Photonic wire bonding: Connecting nanophotonic circuits acorss chip boundaries", *in Proceedings of MOEMS-MEMS SPIE Photonics West* (MOEMS-MEMS-Spie'13), San Francisco (CA), USA, Feb. 2-7, 2013, paper 8613-31, **[invited]**

[C14] C. Koos, J. Leuthold, W. Freude, L. Alloatti, D. Korn, R. Palmer, M. Lauermann, N. Lindenmann, S. Koeber, J. Pfeifle, P.C. Schindler, D. Hillerkuss, R. Schmogrow, „Silicon-organic hybrid integration and photonic wire bonding: technologies for Terabit/s interconnects", *in Joint Symposium on Opto- andMicroelectronic Dvices and Circuits* (SODC2012), Hangzhou, China, Sept. 24-27, 2012 **[invited]**

[C15] N. Lindenmann, G. Balthasar, J. Leuthold, W. Freude, C. Koos, „Broadband low-loss interconnects enabled by photonic wire bonding", *in Proceedings of IEEE Optical Interconnects Conference* (IEEE 2012), Santa Fe (NM), USA, paper WC1, **[invited]**

[C16] N. Lindenmann, G. Balthasar, M. Jorden, D. Hillerkuss, R. Schmogrow, W. Freude, J. Leuthold, C. Koos, "Low-loss photonic wire bond interconnects enabling 5 TBit/s data transmission," *in Optical Fiber Communication Conference* (Optical Society of America 2012), Los Angeles (CA), USA, March 2012, paper OW4I.4

[C17] C. Koos, J. Leuthold, W. Freude, L. Alloatti, D. Korn, N. Lindenmann, R. Palmer, J. Pfeifle, D. Hillerkuss, R. Schmogrow, „Silicon nanophotonics and photonic wire bonding: Technologies for Terabit/s interconnects" *in 6th ITG Workshop on Photonic Integration*, Berlin, Germany, March 20, 2012 **[invited]**

[C18] W. Freude, L. Alloatti, A. Melikyan, R. Palmer, D. Korn, N. Lindenmann, T. Vallaitis, D. Hillerkuss, J. Li, A. Barklund, R. Dinu, J. Wieland, M. Fournier, J. Fedeli, S. Walheim, P.Leufke, S. Ulrich, J. Ye, P. Vincze, H. Hahn, H. Yu, W. Bogaerts, P. Dumont, R. Baets, B. Breiten, F. Diederich, M. Beels, I. Biaggio, Th. Schimmel, C. Koos, J. Leuthold: "Nonlinear optics on the silicon platform" in *Optical Fiber Communication Conference* (Optical Society of America 2012), Los Angeles (CA), USA, March 2012, paper OTh3H. **[invited]**

[C19] P: Schindler, H. Fischer, H. Haisch, N. Lindenmann, A. Alloatti, R. Palmer, D. Korn, D. Hillerkuss, R. Schmogrow. A. Gerster, W. Freude, J. Leuthold, C. Koos, "Optical technologies for off-chip and on-chip interconnects" in *Conference DesignCon*, Santa Clara (CA), USA, January 2012, paper TP5

[C20] N. Lindenmann, G. Balthasar, R. Palmer, S. Schuele, J. Leuthold, W. Freude, and C. Koos, "Photonic wire bonding for single-mode chip-to-chip interconnects", in *Proceedings of IEEE International Conference on Group IV Photonics*, (IEEE, 2011), London, England, September 2011, paper 380-382

[C21] N. Lindenmann, I. Kaiser, G. Balthasar, R. Bonk, D. Hillerkuss, W. Freude, J. Leuthold, C. Koos, "Photonic waveguide bonds - a novel concept for chip-to-chip interconnects", in *Optical Fiber Communications Conference* (Optical Society of America, 2011), Los Angeles (CA), USA, March 2011, Postdeadline paper PDPC1.

[C22] J. Leuthold, W. Freude, C. Koos, A. Melikyan, N. Lindenmann, "A surface plasmon polariton absorption modulator" in *13th Intern. Conf. on Transparent Optical Networks* (2011), Stockholm, Sweden, June 26-30, 2011 **[invited]**

[C23] A. Melikyan, T. Vallaitis, N. Lindenmann, Th. Schimmel. W. Freude, J. Leuthold, "A surface plasmon polariton absorption modulator" in *Proceedings of Conf. on Lasers and Electro-Optics* (2010), San Jose (CA), USA, May 2010, paper JThE77

[C24] T. Vallaitis, D. Hillerkuss, J. Li, R. Bonk, N. Lindenmann, P. Dumon, R. Baets, M. Scimeca, I. Biaggio, F. Diederich, C. Koos, W. Freude, J. Leuthold, „All-optical wavelength conversion using cross-phase modulation at 42.7 Gbit/s in silicon-organic hybrid (SOH) waveguides" in *Procroceedings of Intern. Conf. on Photonics in Switching* (2009), Pisa, Italy, September 15-19, 2009, Postdeadline paper PDP3

[C25] N. Lindenmann, T. Vallaitis, R. Bonk, C. Koos, W. Freude, J. Leuthold, "Amplitude and phase dynamics in silicon compatible waveguides with highest Kerr nonlinearities" AMOP Spring Meeting, Deutsche Physikalische Gesellschaft 2009, Hamburg, Germany, March 2-6, 2009, paper Q57.4